AF885893

Sustainable Textiles: Production, Processing, Manufacturing & Chemistry

Series Editor

Subramanian Senthilkannan Muthu, Head of Sustainability, SgT and API, Kowloon, Hong Kong

This series aims to address all issues related to sustainability through the lifecycles of textiles from manufacturing to consumer behavior through sustainable disposal. Potential topics include but are not limited to:

Environmental Footprints of Textile manufacturing; Environmental Life Cycle Assessment of Textile production; Environmental impact models of Textiles and Clothing Supply Chain; Clothing Supply Chain Sustainability; Carbon, energy and water footprints of textile products and in the clothing manufacturing chain; Functional life and reusability of textile products; Biodegradable textile products and the assessment of biodegradability; Waste management in textile industry; Pollution abatement in textile sector; Recycled textile materials and the evaluation of recycling; Consumer behavior in Sustainable Textiles; Eco-design in Clothing & Apparels; Sustainable polymers & fibers in Textiles; Sustainable waste water treatments in Textile manufacturing; Sustainable Textile Chemicals in Textile manufacturing.

Innovative fibres, processes, methods and technologies for Sustainable textiles; Development of sustainable, eco-friendly textile products and processes; Environmental standards for textile industry; Modelling of environmental impacts of textile products; Green Chemistry, clean technology and their applications to textiles and clothing sector; Eco-production of Apparels, Energy and Water Efficient textiles.

Sustainable Smart textiles & polymers, Sustainable Nano fibers and Textiles; Sustainable Innovations in Textile Chemistry & Manufacturing; Circular Economy, Advances in Sustainable Textiles Manufacturing; Sustainable Luxury & Craftsmanship; Zero Waste Textiles.

More information about this series at http://www.springer.com/series/16490

Subramanian Senthilkannan Muthu · Ali Khadir
Editors

Novel Materials for Dye-containing Wastewater Treatment

Springer

Editors
Subramanian Senthilkannan Muthu
Sustainability, SgT and API
Kowloon, Hong Kong

Ali Khadir
Iran University of Science and Technology
Teheran, Iran

ISSN 2662-7108 ISSN 2662-7116 (electronic)
Sustainable Textiles: Production, Processing, Manufacturing & Chemistry
ISBN 978-981-16-2891-7 ISBN 978-981-16-2892-4 (eBook)
https://doi.org/10.1007/978-981-16-2892-4

© The Editor(s) (if applicable) and The Author(s), under exclusive license to Springer Nature Singapore Pte Ltd. 2021
This work is subject to copyright. All rights are solely and exclusively licensed by the Publisher, whether the whole or part of the material is concerned, specifically the rights of translation, reprinting, reuse of illustrations, recitation, broadcasting, reproduction on microfilms or in any other physical way, and transmission or information storage and retrieval, electronic adaptation, computer software, or by similar or dissimilar methodology now known or hereafter developed.
The use of general descriptive names, registered names, trademarks, service marks, etc. in this publication does not imply, even in the absence of a specific statement, that such names are exempt from the relevant protective laws and regulations and therefore free for general use.
The publisher, the authors and the editors are safe to assume that the advice and information in this book are believed to be true and accurate at the date of publication. Neither the publisher nor the authors or the editors give a warranty, expressed or implied, with respect to the material contained herein or for any errors or omissions that may have been made. The publisher remains neutral with regard to jurisdictional claims in published maps and institutional affiliations.

This Springer imprint is published by the registered company Springer Nature Singapore Pte Ltd.
The registered company address is: 152 Beach Road, #21-01/04 Gateway East, Singapore 189721, Singapore

Contents

About the Editors

Dr. Subramanian Senthilkannan Muthu currently works for SgT Group as Head of Sustainability, and is based out of Hong Kong. He earned his Ph.D. from The Hong Kong Polytechnic University, and is a renowned expert in the areas of Environmental Sustainability in Textiles & Clothing Supply Chain, Product Life Cycle Assessment (LCA) and Product Carbon Footprint Assessment (PCF) in various industrial sectors. He has five years of industrial experience in textile manufacturing, research and development and textile testing and over a decade's of experience in life cycle assessment (LCA), carbon and ecological footprints assessment of various consumer products. He has published more than 100 research publications, written numerous book chapters and authored/edited over 100 books in the areas of Carbon Footprint, Recycling, Environmental Assessment and Environmental Sustainability.

Ali Khadir is an environmental engineer and a member of the Young R^4, including Journal of Environmental Chemical Engineering and International Journal of Biological Macromolecules. He also has been the reviewer of journals and international conferences. His research interests center on emerging pollutants, dyes and pharmaceuticals in aquatic media, advanced water and wastewater remediation techniques and technology. At present, he is editing other books for Springer in the field of nanocomposites, advanced materials, and the remediation of dye—containing wastewaters.

Dye Pollution in Water and Wastewater

Karishma Maheshwari, Madhu Agrawal, and A. B. Gupta

Abstract The presence of dye in water stream leads to unexceptional effects on living life. As dyes are consumed globally from small-scale to large-scale industries inculcating tanneries, food, cosmetic, textile, medicinal sectors with the production of 1,000,000 tons all over the world. Majorly, the textile sector plays a pertinent role in dye emissions into the ecosystem. Only dying industries discharge about 7.5 metric tons annually. The complex structures of dye comprising of aromatic rings bonded with different functional groups having π-electron could absorb light within 380–700 nm spectra. They impart coloration due to the presence of chromogens and chromophores. Out of several natural and synthetic dyes, azo group proliferation has been highly carcinogenic due to amine and benzidine emissions. Besides this, the fact of being non-biodegradable makes the dye molecules last longer in the environment producing hazards. Henceforth, eradication of dye molecules from wastewater is thus needed before discharging the stream into the environment with long- and short-term effects. The severe implications have been reported for the aquatic life due to direct contact. Whereas to human life, there were observations from skin irritations to cancer-like disease. Several approaches were reported for the treatment of dye-containing streams but the exploration for the best available technique is still ongoing. So, this chapter collects comprehensive facts about dyes, its harmful effect, and approaches for the treatment at a worldwide level and Indian perspective. Additionally, the article draws out the comparative analysis for available techniques and states the recent advancements for the purification of dye-containing wastewater.

Keywords Textile · Pollution · Toxicity · Effects · Treatment

---

K. Maheshwari · M. Agrawal (✉)
Department of Chemical Engineering, Malaviya National Institute of Technology Jaipur, Jaipur 302017, India
e-mail: magarwal.chem@mnit.ac.in

A. B. Gupta
Department of Civil Engineering, Malaviya National Institute of Technology Jaipur, Jaipur 302017, India

© The Author(s), under exclusive license to Springer Nature Singapore Pte Ltd. 2021
S. S. Muthu and A. Khadir (eds.), *Novel Materials for Dye-containing Wastewater Treatment*, Sustainable Textiles: Production, Processing, Manufacturing & Chemistry,
https://doi.org/10.1007/978-981-16-2892-4_1

1 Introduction

Water is the basic amenity for the existence of living bodies [1]. Humans in the sake of historical cultivation and globalization have grown themselves in terms of technologies [2]. This industrialisation has led to severe water contamination leaving a bad impression towards environmental health especially aquatic bodies [1]. For instance, dyes are generally used to impart colour whether in textile, food, cosmetics, medicinal, etc., and therefore, the resulting water has a significantly strong percentage of dye wherein 50% of it has been lost in the effluent stream [3]. The waste stream has been utilized for washing, bathing, etc., or in the majority cases the stream has been discarded into rivers, lakes, canal, and drainage [4, 5]. The assessment of the quantity of dye in waste stream is a pertinent examination to be performed prior to the disposal as even a small quantity of 1 mg/L could result in coloration and an undesirable concentration for consumption [6]. Majorly, dyes are organic multifaceted molecules and have the property of being attached to several surfaces like fabrics, leathers, etc. In accordance to the recent survey, there have been 1,00,000 commercially available dyes used all over the globe with a record of 10,000 tonnes per year of consumption [7]. With this huge squandering of textile dyes, the fact that effluent being highly complex in nature makes the disruption of an ecological system [5, 8].

Textile discharges may comprise of several pollutants coupled with organic loads. Pollutants may include highly complex dye structures, heavy metals, salts, surfactants, chlorinated chemical species, and auxiliary compounds, etc. [9]. Out from which a comprehensive study by Al Prol (2019) revealed that a major part of the stream (93%) of the feed to the textile industries emerges out as the pigmented wastewater and the remaining comprises of other chemical complexes. Moreover, it has been a difficult task to quantify the exact concentrations of toxic chemicals lost during the process due to the plethora of substances [10]. Hassaan (2017) provided the fact that the production of 1 kg of textile cloth approximately 0.15 m^3 of water is utilized with the statistics of 3000 m^3 discharge processing 20 tons of fabrics per day [11].

Due to the discharge being rich in several chemicals, every pollutant has its toxicity level and imparts threat accordingly. But almost all chemicals in the discharge stream are non-biodegradable in nature with carcinogenic properties [12]. The existence of dye molecules in the hydrosphere reduces the penetration of sunlight deep into the water bodies and thereby inducing annoyance in the biological pursuit of aquatic life [13]. There have been toxicological outcomes like growth inhibitions, reduced ingestion capacity, accumulation in living cells, increased enzymatic activities, reduced reproductive potential kidney dysfunctions, protein level decrements, respiratory problems, opercular movements, histopathological variations, sub-lethal consequences, etc. Not only these mentioned but many more harmful effects were reported due to toxic pollutants to the aquatic and living life [14]. But main characteristics depend on the analysis of every component in the discharge, which is the main step before opting the method for treatment. Therefore, many investigations were performed for analysing the long-chain chemical elements in the waste [15]. So, it could be addressed that the treatment proposed maybe location-specific in nature. Sources of emissions are firstly to be identified. Sources could be varying as it can fluctuate from the domestic level to the industrial level. The functionally active chemical core species which inflate the real characteristics of water is the chromophore.

There were several divisions based on chromophore namely, azo, cyanine, xanthene, nitro, quinone-imine, indigoid, acridine, oxazine, anthraquinone, triaryl-methane, phthalein, triphenylmethane dye, diarylmethane, etc. Chromophore is the nimble organic parts portraying the spatial atoms localization and the adsorption of electromagnetic waves is due to the excited state of molecules [16, 17]. Moreover, the preceding section describes about the classification based on chromophore in detail with their important characteristics.

2 Classification of Dye

There were several bifurcations of dyes depending on chemical structure, ionic charge, etc. Mainly these two were explained briefly in the articles. Simply these are colour imposing elements to the fabric which connect its molecule to the surface of basic cloth. So, the classification based on surface charge includes ionic dyes and non-ionic dyes. Whereas structural division includes azo, cyanine, xanthene, nitro, quinone-imine, indigoid, acridine, oxazine, anthraquinone, triaryl-methane, phthalein, triphenylmethane dye, diarylmethane, etc. depending on chromophore [14]. The preceding sub-section included the classification of dyes and their chemistry.

2.1 Classification of Dyes Based on Charge

Surface charge-based classification of dyes includes ionic and non-ionic bifurcations wherein the ionic dyes include cationic and anionic which are mostly used category of the dye. Moreover, anionic dyes have their wide application due to the huge consumption in textiles due to the affinity towards the fabric in forming covalent bonds and these dyes are brighter dying agents in comparison to other dyes [18]. Therefore, anionic dyes are generally applied in greater quantity. Whereas the cationic dyes are although water-soluble, but they inhibit some biological activity due to the positive surface charge and chemical property used in lesser quantity as they possess benign coloration and malignant tumours. Cationic dyes include cyanine, azine, acridine, azo, anthraquinone, xanthene, oxazine, etc., and some of the anionic dyes namely, nitroso, nitro, triphenylmethane, anthraquinone, azo groups, etc. [15]. Both anionic and cationic dyes are carcinogenic dyes, but the hazards associated with the anionic

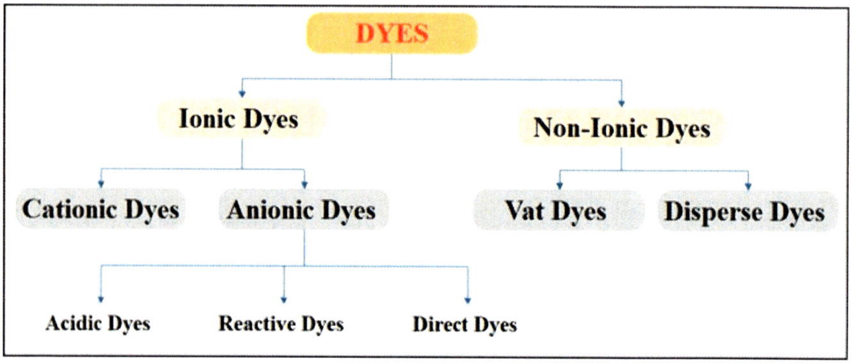

Fig. 1 Classification of dye based on charge present on surface of molecule

dyes have more mutagenic effects like bladder cancer, allergies, dermatitis, asthma, etc.

Additional to the ionic division, dyes could be classified as non-ionic dyes. Such non-ionic dyes are the neutrally charged surface molecules. It includes vat and disperse dyes in which vat dyes react with leuco salts after getting reduced in alkaline medium whereas disperse dyes are insoluble in water and maybe form hydrophobic liquid dispersion [19]. Vat dyes have very less toxicity whereas disperse dyes are relatively carcinogenic in nature [20]. Non-ionic dyes include indigoids, anthraquinone, nitro, azo, styryl, etc., groups associated with the chemical alignment. The classification of dyes based on the charge is as shown in Fig. 1.

2.2 *Classification Based on Chemical Structures*

Dyes are basically the organic chemical structures with three important groups, i.e., (a) chromophore, (b) auxochrome, and (c) matrix in the molecular structure. Out of which chromophore is the essential spatial group is the collection of atoms commonly comprising of nitro, azo, nitroso, carbonyl, alkenes, etc. [9]. Dye entity encompassing these structures becomes chromogenic moreover these when added with auxochrome enhances the dyeing abilities of the molecule. Auxochrome part might be acidic or basic in nature, i.e., hydroxide, sulphite, carboxylic, etc., are categorized as acidic while amino, alkanol amine, etc., are basic ones. Remaining is the part named as matrix. Each type of classification has an attached group to the molecular structure and has its own colour index reflecting the corresponding chronological order as tabulated in Table 1 [21]. Dye effluent with functional group is lesser likely to be treated with the conventional techniques. Moreover, the type of functional elements in the effluents has been complex in nature due to the several attached chemical components. The chromophore-based division has been detailed in Table 1 with the assigned colour indexed ranges as reported in the literature. Acridine is the

Table 1 Chromophore based proliferation of dyes [14]

Class of dye	Chromophore structure	Colour index ranges
Acridine dyes	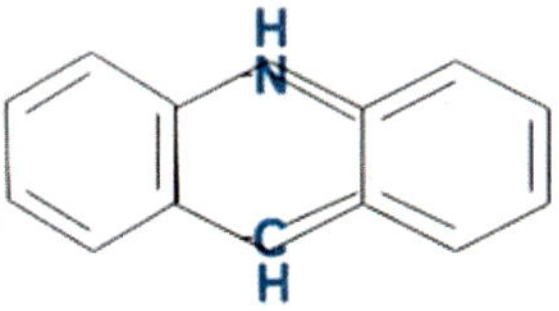	46,000–46,999
Anthraquinone dyes	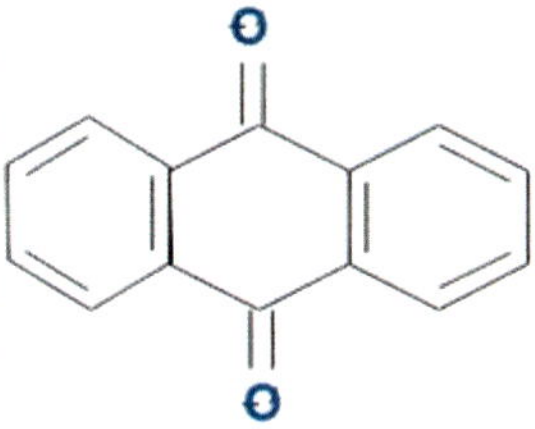	58,000–72,999
Azo dyes		11,000–39,999
Cyanine dyes	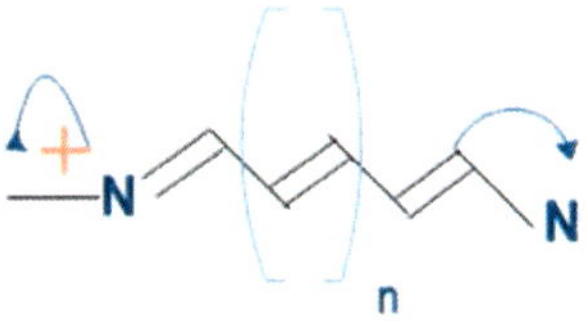	74,000–74,999
Diarylmethane dyes	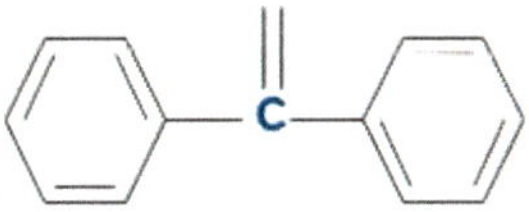	42,000–44,999
Indigoid dyes	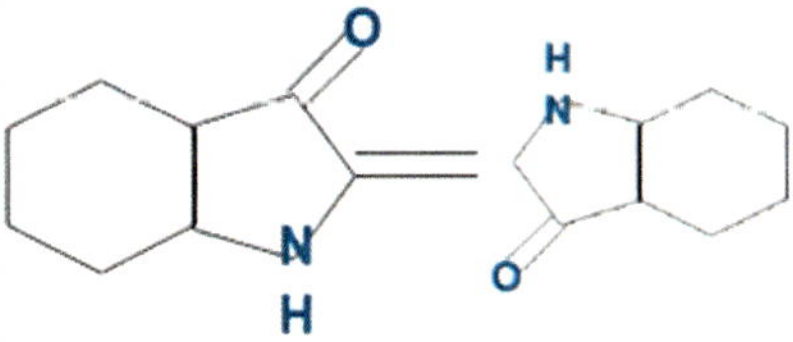	73,000–73,999
Nitro dyes	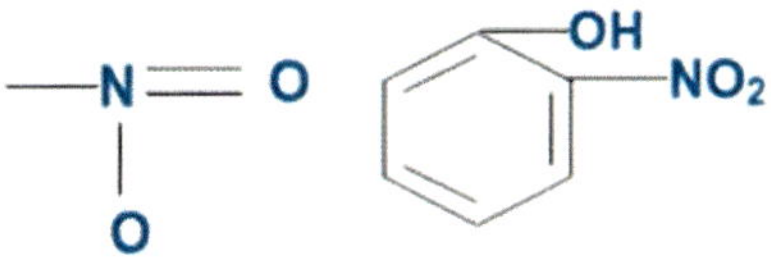	10,300–10,999
Nitroso dyes		10,000–10,299

(continued)

Table 1 (continued)

Class of dye	Chromophore structure	Colour index ranges
Oxazine dyes	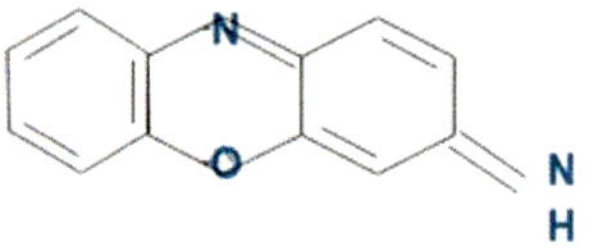	51,000–51,999
Phthalein dyes	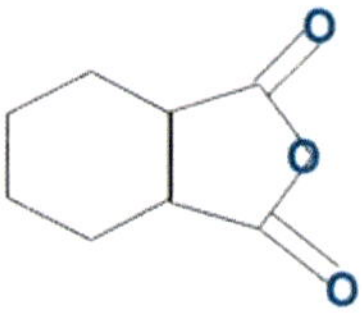	52,000–52,999
Quinione-imine dyes	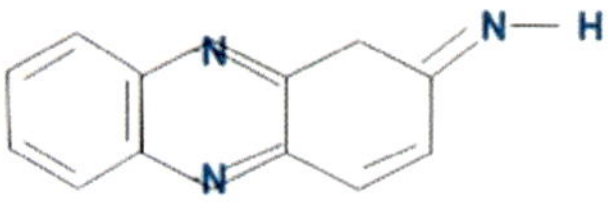	77,000–77,999
Triarylmethane dyes	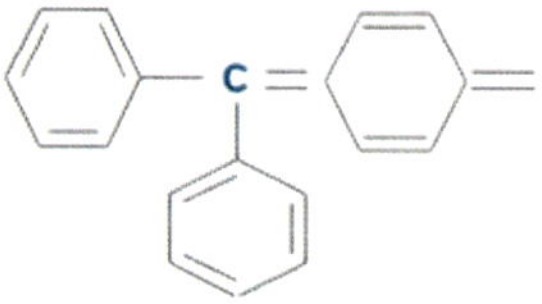	42,000–44,999
Triphenylmethane dyes	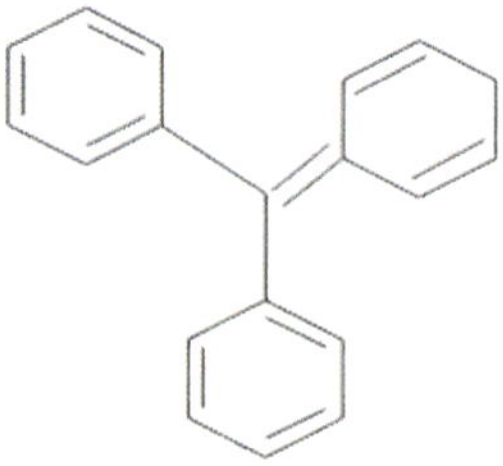	41,000–41,999
Xanthene dyes	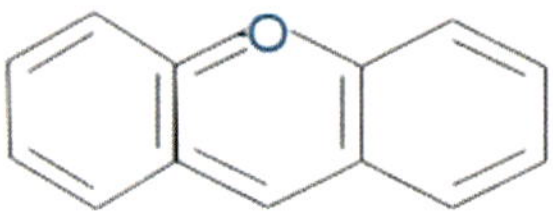	45,000–45,999

chemical structure of dye based on the organic matrix with nitrogen-derived heterocyclic formula. Earlier, these were the dyes playing important once at a time but have niche application as far as the present scenario is concerned [6]. It has a basic nature and colourless crystal-like appearance. Some of the examples include pyridine, quinoline, etc., having C-H bonds located centrally in the structural formula. Anthraquinone relates to the abundant type of dye which is usually colourless but turns blue to red by incorporating electron donor groups like amino or hydroxy at the 1, 4, 5 or 8th position. They can occur naturally or may be categorised as synthetic dyes. Such dye comes under indigo stuff. They are generally dispersed and reactive dyes. Azo dyes are the chemical structures with linkage of R-N=N-R' aryl nature which are consumed in textiles, leather, foods, paper field. Mostly, methyl orange, disperse orange 1, phenolic diazo, etc., are widely used in these dyes. Cyanine dyes are synthetic dyes which is a heterocyclic system wherein the structure has nitrogen conjugated with pyridine, imidazole, quinoline, thiazole, pyrrole, etc. [22]. This is highly consumed in biotechnology for analysis, biomedical purposes, and labelling.

Triarylmethane dye comes under synthetic chemical dyes comprising of triphenylmethane as their supporting molecule. These are immensely coloured chemicals, for instance, crystal violet is one of the triarylmethane dye. Widely implemented for biotechnological applications containing nitrogen in the end so that it can be linked to either nucleic acid or protein molecule [23]. Indigoid dye is distinct organic species with blue colour and was naturally extracted from plants. However, they may be synthetically obtained and used for colouring denim fabric and blue cloths. Nitro group is an aromatic dye whose structure contains NO_2, imino, hydroxyl, OH, etc. linkages. Sometimes may consists of chloride, COOH, etc., substituents. Another group, Nitroso has NO linkages in the chemical structure in the organic moiety. There may be nitrosoalkanes, nitrosothiols, nitroamines, alkyl nitrites, etc. Oxazines are heterocyclic structures with one nitrogen and oxygen atoms attached to double unsaturated 6 member ring-like structures. It consists of ifosfamide, morpholine, etc. Phthalein are basically the indicators for pH due to the capability to change colour. Majorly formed of phthalic anhydride with conjugated phenols [24].

Quinone imine dyes are the results of oxidation of para-aminophenols and condensation of paranitrosophenols. It comprises of sulphur, nitrogen, oxygen atoms with heterocyclic structure and used for the production of thiazine, azine, oxazine, respectively. Xanthene is the organic yellowish solid-like structure commonly soluble in organic solvents and is obscure, but its derivatives are still used [25]. The direct emission of molecules is the major topic of concern which is delineated in the preceding section.

3 Sources of Dye in Aquatic Life and Environment

Emerging of the polluting molecules is a major point to be analysed as according to the status of production and consumption of synthetic and natural dyes in the field of textiles, food, tanneries, etc. Majorly these are the highly detected areas from where ample wastewater has been discharged. However, there is no such criterion defining discharged by direct or indirect ways. But the reported analysis of states that India emits out about 9000 m^3 of an effluent stream without treating it. Some major sources have been delineated in Fig. 1 as textile, industries, chemicals, food, tannery, household, printing, etc. [26].

Textile has notably been the major origin for dye effluent as they have dyeing steps and it discharges direct into the water bodies. For instance, dyeing the cellulose fabrics after which, the stream discharged contains 10–40% sulphur dyes 10–50%

of reactive type dyes. Statistical data with textile dyes states that 200 billion l of pigments are produced annually and up to 50% are the effluents dispensed directly in the aquatic system. Every year it has been estimated that 2,80,000 t of textile dyes are lost in this sector. As of now, 20% of industrial pollution has been reported due to textile emissions. Another dyeing disposal point is the household hair dyeing, dye-containing cosmetics, household chemicals, which during usage is mixed with water and disposed at the end to rivers, sewages, etc. In an investigation researcher evaluated wastewater from 17 flats consuming 750 L capacity of water. The study revealed that there were 11 synthetic dyes with acidic orange 7, acidic orange 24, and acid blue 25. During the period of analysis people were asked not to use dyes for hair or clothes so that clear statistics could be obtained for dyes presence in household chemicals [27].

The food industry is another dye dispensing source. The fish industry is the major unauthorized point leading to contamination of water bodies as the farms discharge dyes like malachite green dyes [14]. Khodabakhshi and Amin (2012) evaluated the fish farm effluent for the presence of malachite green and reported about 0.006 to 0.4 μg/l concentration of dye in the effluent [28]. Moreover, malachite green in fishing industries effluents is ranged from 265 to 1663 μg/kg revealing an aqua system has been directly hindered by the presence of dye due to the untreated discharges from food industries. Also, the colorants used in food industries are a major source for dye emissions in the water. Currently, it was found that at least 15 dyes out of 45 were authorized and others were used as colorants [14]. Li et al. (2013) explored the colorants usage in the food industry and found that about eight dyes were used in food industries illegally which are banned [29]. Such banned dyes were utilized in textile sectors and were found to be used for food sectors. In an investigatory article, Rahi 2018 collected the effluent from Sanganer textile industrial region from Rajasthan and reported the huge TDS of about 3000–3500 ppm and TSS 5000–7000 ppm [30]. Another research states that Ahmed et al. (2020) evaluates the physicochemical properties from the collected samples discharged from the textile effluent reporting the turbidity high up to 250 FTU and high electrical conductivity of 1300 μs/cm with heavy metals like lead (0.5 ppm), nickel (0.2 ppm), and copper (0.11 ppm) [31]. Moreover, there were huge total dissolved solids found in the sample analysed by Khatun et al. (2020) states that textile effluents collected from the discharge end of pakiza dyeing and printing industries and reported high COD (400 ppm), TDS (2100 ppm), and TSS (200 ppm) [32]. Also, there were investigations carried out in which huge organic matters within the range of 1–5 %, electrical conductivity of 13000 μS/cm, and high sulphate content of 4 mg/kg [33]. Cases like this shows that without proper treatment of effluents the sectors like food, pharmaceutical, fishing, textile, etc., make a direct source for discharges into aquaculture [34] (Fig. 2).

4 Textile Dye Emissions

There were various case studies, investigations, researches, and project established for obtaining the calculative data for textile industries and their emissions. However, there is no such obligation taken after them to pre-treat the discharge before disposal. The effluent stream might be subjected to have high organic matter, coloration, heavy metals, turbidity, hardness, dissolved solids, suspended solids, dissolved oxygen, ionic species like sodium, potassium, calcium, magnesium, chloride, sulphate, nitrate, fluoride, etc., and alkalinity in most of the cases [35]. However, there might be differences in composition as the chemical and physical properties varies based on the location, type of industries, and quality of effluent being discharged [36]. This section provides an overview of the global situations and national-level status of textile industries and their emissions. Preliminary case studies evaluated by researchers investigating the physicochemical properties of textile effluent discharged from various industries have been tabulated in Table 2.

4.1 Worldwide Level

Globally, the overall emissions reported for textile effluents are immensely high. It has been already highlighted that 7×10^5 tons of dyes have been consumed. For instance, even if 10% out of which gets discharged would create a huge negative impact on living and aquaculture [4]. One of the leading producers and consumers of synthetic azo dyes are India, China, Malaysia, and Germany. One report by Madamwar et al. (2019) states that Malaysia holds about 1.3% of the world's contribution for synthetic dyes production and Germany also has the main hand on for about 1.1% contribution [43]. Moreover, China holds about 36% of the U.S. market for textile imports and tops the position for sharing 29% of the European Union textile imports in the world [44]. One of the reports by Samir et al. (2021), states that samples collected from textile industries situated in Zhenjiang, China have azo, reactive, disperse, and sulfonic dyes in the composition [45]. Another focus highlighted by the researcher performed assessment of the discharge from Tunisian textile industries reported high characteristics like BOD (552–900.7 mg/L), COD (1595.7–2104 mg L), TSS (252.33–604.67 mg/L), turbidity (255.67– 549.67 NTU), and conductivity (9.69–13.89 mS/cm) [46]. Moreover, Li et al. (2018) evaluated the composition of synthetic dyeing industries revealing disperse dyes, additives, acetates, soluble micropollutants [47]. The emissions all over the world are found increasing at a rapid rate with major sources as illustrated in Fig. 2. The year-wise emission and the extrapolations till 2023 for dye emissions are portrayed in Fig. 3.

Table 2 Effluent discharges from various industries

Discharge point	TDS (mg/L)/Electrical conductivity (μs/cm)	TSS (mg/L)	pH	BOD (mg/L)	COD (mg/L)	Ionic distribution	References
Taloja industrial belt of Mumbai, India	12,023.6	13,499.2	6.5–8.5	776.2	4206.2	Chloride-260 ppm	[37]
Industrial effluent at Bhairavgarh Ujjain city, India	418		7.8–8.7	–	–	Calcium-270 mg/kg, Sodium-80 ppm, Potassium-60 ppm, Chloride-669 ppm, Magnesium-121 ppm	[35]
Shivasakthi Textile Processors effluent discharge, Tirupur, India	4280	47	9.76	80	317	Calcium-272 ppm, Sulphate-75 ppm, Chloride-1912 ppm, Sodium-1600 ppm, Potassium-38 ppm	[38]
Dye waste site at Binjhole, Haryana, India	4,090	–	7.5–8.2	33	734	Sulphate-3 mg/kg, Nitrate-1.66 mg/kg, Phosphate-11.17 mg/kg	[39]
Dyeing industries discharge, Ludhiana Punjab, India	2524.56	46.90	6	66.75	0.03	Colour-674	[40]
Industrial effluent from Jajmau, Kanpur, India	2286	–	7.8	–	–	Calcium-261 ppm, Magnesium-134 ppm, Chloride-683 ppm, Nitrate-76 ppm, Sulphate-265 ppm, Fluoride-2.6 ppm, Arsenic-0.004 ppm, Chromium-0.02 ppm, Lead-0.001 ppm	[41]

(continued)

Table 2 (continued)

Discharge point	TDS (mg/L)/Electrical conductivity (μs/cm)	TSS (mg/L)	pH	BOD (mg/L)	COD (mg/L)	Ionic distribution	References
Narsingdi, Rajshahi and Gazipur Districts of Bangladesh	3000–3700 ppm	400	7.5–8.4	270	784	Bicarbonates-360 ppm, Chloride-654 ppm, Sulphate-448 ppm	[42]

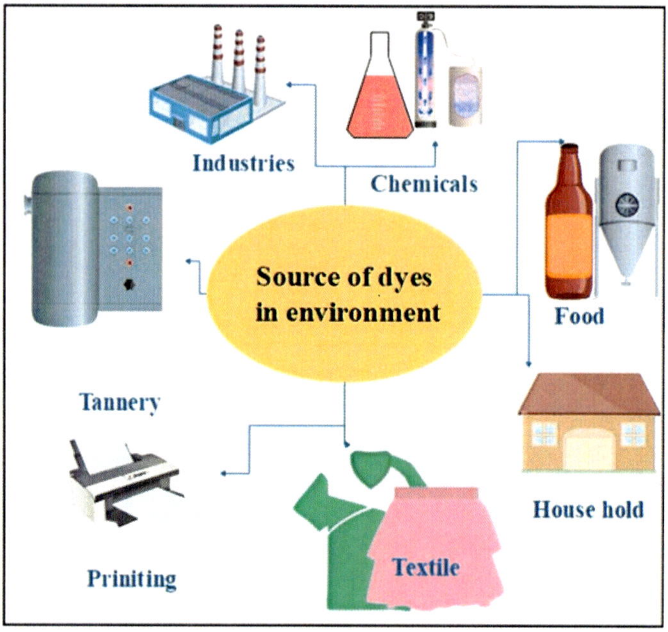

Fig. 2 Sources of dye in environment [24]

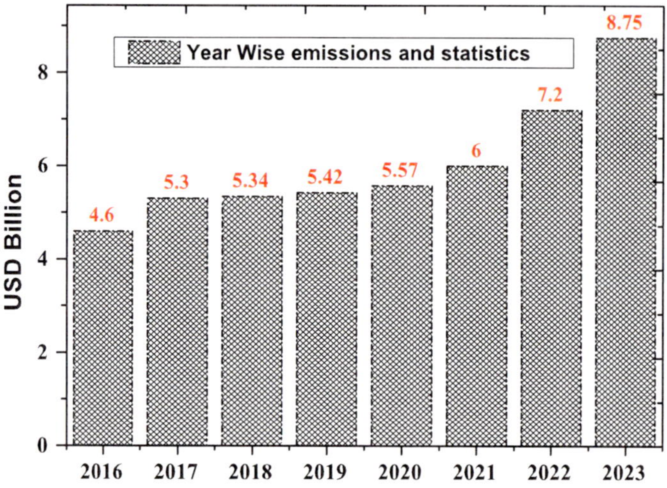

Fig. 3 Worldwide dye effluent emissions and the future extrapolations

4.2 Indian Scenario

As far as the Indian scenario is considered there are wide varieties of dyes utilized in fabrics, food, and pharmaceuticals. Most importantly the textile sector has the highest consumption of dyes in India. The state-wise distribution has been evaluated and provided in Fig. 4 till date [43]. However, to the Rajasthan's ground level statistics there have been basically textile factories from small scale to the industrial level. Due to the cultural heritage of states value, Rajasthan is famous for the cloth market and prints. It was evaluated that the states consume 30,000 tons of cotton and polyester for dyeing purposes. Major areas concerned to the textile industries are Pali, Balotra, Bagru, Sanganer, etc. [2]. Rathore (2012) studied the loading pollution caused in dyeing industries in Rajasthan's Pali area who stated that it has been the biggest area of the state with 989 small-scale industries for dyeing of fabrics. The industries generate about 13 MLD liquid wastewater whereas solid waste of 10 tons/day [48]. Moreover, Sharma et al. (2014) studied the influence of effluent discharged into water for physicochemical properties and reported that liquid waste is about 18 MLD and the solid waste emissions were 8–10 tons/day [49]. Bagru region of Rajasthan has about 250 textile units and 600 dyeings were there in Sanganer area [48]. Ludhiana another region has about 268 industries out of which 10 are running at large-scale level and the remaining are small level dyeing sectors. These textile industries are generating about 150 MLD of an effluent stream consisting of organic as well as inorganic pollutants [40] (Fig. 4).

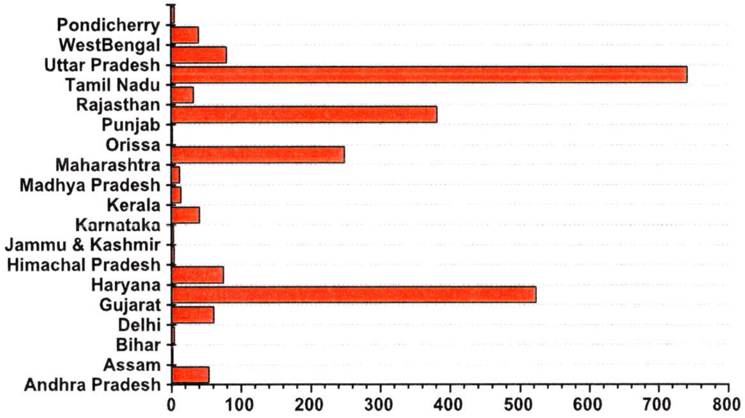

Fig. 4 Textile industries distribution in India [44]

5 Toxicity Associated with Textile Effluents

Toxicity can be justified with the amount of the substance being poisonous. It is the by which the organism, environment, cell, tissues, etc., gets affected. However, the effect on every individual may vary [50]. An investigator Novotný et al. (2006) reported a comparative assessment for toxicity of two azo and anthraquinone dyes on the living organism like bacteria, ciliates, and microalga. Amongst the dyes tested, EC50 for disperse blue 3 was maximum i.e. 0.5 mg/l (Selenastrum capricornutum) relative to Vibrio fischeri with 488 mg/l EC50 [51]. Insights to the measurements revealed that algal organisms showed higher sensitivity to the dye tested than that depicted by bacteria. A briefed toxicological analysis over microorganism with the class of dyes have been provided in Table 3.

Not only aquaculture suffers from the toxicity of the direct untreated discharges but the human beings, animals, plants whosoever comes in exposure with the dyes get affected. Human's contact with azo dyes especially is harmful as it leads to carcinogenic effects. There may be other effects like inhalation issues, bad intestinal activities, respiratory problems, etc. Chung (2016) explains the effect of azo dyes on human health in his article which states that aromatic amines attached to the conjugated dye structures lead to disorders in urinary tracks, hematopoietic, porphyria, hypertension, sensitivity, etc. Not only this but the contact can also leads to immune deficiency and leading to autoimmune human body [57]. Feng et al. (2012) delineated the intestinal effects caused by azo dyes briefing that ingestion of azo dyes can cause hepatocarcinogenicity in animal mice and mark ably leading to carcinogenetic effects in humans [58]. An investigator Chequer et al. (2009) evaluated the mutagenic activity of disperse dyes red 1 and orange 1 due to the immense effect of these dyes on human lymphocytes they mentioned that presence increased the micronucleus frequency in the human bodies [59]. Sathishkumar et al. (2019) explained that textile effluents consisting of highly hazardous dyes like Congo red may inhibit the effects till death [60]. As Congo red is an azo dye with one sulfonic, hydroxyl, and chromophore group and combination of all the three makes the chemical species highly toxic. And therefore, the proper treatment of such effluents is needed [2].

6 Available Treatment Method and Recent Advances

Various reported approaches governed by different mechanisms have been studied and applied for dyes and textile water treatment. There were membrane-based techniques, temperature-driven processes, electric potential-based approaches, and sorption techniques reported in the literature for dye remediation [61]. However, every technology explored had its own advantages and disadvantages. So, for the same, this section draws out the capabilities of several technologies reported in the literature with their critical aspects. Considering the process operations, energy consumption, economics, efficacies, and process parameters for recent dye removal investigations.

Table 3 Class of dyes and their toxicological effects

Dye class	Dye and its toxicity(mg/l) (EC50/LC50)	Species and taxon	Effects	References
Azo	Acid orange 7 (16), Food yellow (22), Reactive black 5 (28), Procion yellow (71), Procion navy (19), Procion crimson (35)	Bacteria (Vibrio fischeri)	Irritability, restlessness, depression	[52]
Anthraquinone	Remazol, Congo red, brilliant blue R, Reactive orange 16, Disperse blue 3	Ciliates (Tetrahymena pyriformis)	Decreased ingestion, inhibition of growth, rounding effect on cell	[51]
Azo, Methane, Azo, Anthraquinone	Reactive red 141 (96), Basic red 14 (11), Reactive black 5 (29), Acid orange 7 (100), Food red 17 (61), Food yellow 3 (72), Vat green 3 (6)	Algae (Chlorella vulgaris Selenastrum Capricornutum)	Cell size decrements, Accumulation in cells	[53, 54]
Anthrone, Anthraquinone, Azine, Azo	Vat blue 20 (96), Vat green 1 (142), Vat green 9 (96), Acid blue 324 (12), Acid blue 80 (124), Acid violet 48 (102), Direct blue 293 (147), Acid black 1 (54), Acid black 24 (67), Acid black 26 (43), Acid black 94 (15), Acid blue 113 (78), Acid brown 235 (6), Acid brown 354 (92), Acid green 111 (72), Acid green 68 (39), Acid orange 33 (66), Acid orange 7 (56), Acid orange 95 (70), Acid red 119 (12), Acid red 131 (85), Acid red 266 (22), Acid red 374 (8), Acid yellow 42 (77), Acid yellow 49 (65), Acid yellow 61 (58), Direct blue 71 (66), Direct green 26 (49), Direct red 227 (126), Direct red 23 (107), Direct red 81 (78), Disperse blue 79 (113), Br Disperse (128), blue 79 Cl (45), Disperse orange 30 (85), Disperse orange 61 (116), Mordant yellow 10 (1), Mordant black 11 (13), Reactive black 5 (135), Reactive red 120 (153), Reactive red 195 (120), Reactive red 83 (11), Reactive yellow 15 (15)	Selenastrum Capricornutum (Algae)	Growth inhibitors, health hazards, Irritability, Restlessness, etc.	[55]

(continued)

Table 3 (continued)

Dye class	Dye and its toxicity(mg/l) (EC50/LC50)	Species and taxon	Effects	References
Azo	Disperse red 1 (1.9), Remazol brillant blue R (No data)	Hydra attenuata (Invertebrates Cnidaria)	Increased enzymatic activity	[56]

Baghel et al. (2018) evaluated the vacuum membrane technology for blue-black naphthol dye revealing a potentially good flux of 54 kg/m^2 h at 85 °C temperature with flow rate of 5 lpm and feed pressure of 750 mm Hg but considering the capital expenditure and the membrane fouling the process lacks for the treatment at industrial scale [62]. Similarly, another investigation by Khumalo et al. (2019) focused to overcome the fouling problems by nanoparticles incorporation and worked at membrane distillation for toxic congo red dye removal resulted in the flux of 0.0041 L/hm^2 but the cost economics were still an unresolvable issue [63]. Moreover, Heidari et al. (2019) implemented a catalytic oxidation process using nanocomposite found the 100% removal of reactive red 198 at a dose of 0.3 g/L and reaction time of 25 min but there were no physicochemical parameters reported for treated water leaching and is not industrially acceptable [64]. Moreover, Mook et al. (2017) implemented electrocoagulation for reactive black 5 treatment and found that an electrolyte dosage of 0.11 g/L and 0.075 A current a removal of 83.5% was obtained. This approach is although energy efficient but the outcome from the process was relatively less [65]. Recent application of adsorption was explored by Nasseh et al. (2020) synthesizing metallic novel iron, silicate, copper-based adsorbent for removal of congo red reporting 100% removal of contaminants at 2 g/L dose and 200 min of contact duration from wastewater [5]. Moreover, based on similar kind of operation advanced technique of capacitive deionization (CDI) is yet a topic for exploration in the field of dye removal.

CDI is a combination of sorption and potential applied process. Till date the investigations were reported for salt sorption, heavy metal remediation, anionic pollutant removal like fluoride, etc. [66]. It has good process feasibility, economics, energy statistics, and outcome so to the best of author's knowledge this could be a plausible approach for dye-containing wastewater treatment. Therefore, a brief understanding of phenomenon and the treatment of wastewater were explored in the chapter within upcoming section.

6.1 Treatment of Dye Emissions via Adsorption

Adsorption is the most employed technology for the treatment of dyes effluents. Enormous investigations were reported in the literature for dye removal due to variety of adsorbents and the ease in operation. There were investigations found using activated carbon, activated carbon derived from biomass, carbon nanotubes, metal–organic frameworks, synthetic adsorbents, nanoparticles, etc. [67]. Selection of adsorbent plays a major role for application as the properties namely specific surface area. Higher the specific surface area more will be the adsorption due to the porous cavities adsorbing pollutants. Moreover, the process includes several parameters which need to be optimized like pH, dosage, temperature, initial concentration, and contact time [68]. Dye molecule specially has affinity towards adsorption on surface of adsorbent due to the charge attraction and Vander wall forces leading to chemical and physical adsorption [69]. The investigations performed by researchers state that adsorbents tend to have the highest removal rates and sorption capacities with ease in operation and economically feasible [70]. Recenet reports from conventional carbon family adsorbents composites and metal–organic framework has gained attention. In this area, a research performed by Liu et al. (2019) revealed that mesoporous carbon-based bimetallic Zn/Cu incorporated adsorbent leads to the uptake capacity of 1000 mg/g for congo red dye removal 100 mg/L and dosage of 0.02 g [71] In the recent investigation, Zolgharnein et al. (2020) states that a novel metal–organic framework of Nickel possessed potentially good adsorption capacity of 242.1 mg/g for congo red dye removal with initial concentration of 50 mg/L, dosage of 0.0107 g and temperature of 45 °C [72]. Moreover, from the carbon family, adsorbents from waste have been explored for textile effluent treatment wherein Maheshwari et al. (2020) attempted the congo red removal from the walnut shell derived biochar using microwave-assisted synthesis and ultrasonic treatment achieving the adsorption capacity of 75.75 mg/g at an adsorbent dose of 0.2 g and consuming the lowest energy in the approach of around 0.7 kWh with an effective cost of Rs. 2.81 and treatment cost of Rs. 4.9 [1]. But the related shortcoming noted in the composites and carbon-based recent examinations were the cost of synthesis and the leaching of metal from adsorbent in the treated water. Whereas considering the biochar-based adsorbents which are lacking the applications at industrial level. Henceforth, the treatment of dye effluents requires a new process for treatment which is cost efficient and has implication in industrial aspects.

6.2 Treatment of Dye Emissions via Coagulation

Treat of dye effluents through coagulation has also gained attention due to low cost, ease in operation, and can be operated without electricity. Dalvand et al. (2016) performed the synthesis of coagulant from Moringa stenopetala seed extract which could achieve 98.5% removal in 240 ppm concentration [73]. Shankar et al. (2017) performed the pre-treatment of textile effluent having and corresponding percentage removal: Acid Red 94 (41.5%), Acid Yellow 1 (27%), Direct Green 26 (26.2%), and Reactive Blue 21 (43.5%) by alum dosing [74]. In the same string, Le (2020) attempted the synthesis of coagulant from mucilage extract obtained from dragon fruit resulted in the turbidity removal of 95% from the dye wastewater [75]. But the problem associated with the coagulant is mixing time distributions and the lesser removal efficiency.

6.3 Treatment of Dye Emissions via Filtration

Membrane filtration has gained enormous horizon in the field of wastewater treatment. For effluent treatment there were nanofiltration, reverse osmosis, ultrafiltration modules applied, and pilot plant assembled at the discharge point of dye industries. Membrane processes depend on the pressure parameters and the pore size of membrane used. From the literature, it has been seen that nanofiltration when applied for the treatment of dye effluent from using thin-film composite achieved a high rejection of 99.99% removal with an initial TDS of 5000 ppm operating at the pressure of 400 psi [76]. Additionally, the Baghel et al. (2018) performed the vacuum membrane distillation for naphthol blue-black removal revealing the 99.7% removal with permeate flux of 53.51 kg/m^{2}h at 85 °C temperature, 5 lpm rate, initial concentration of 30 ppm and a pressure of 750 mmHg [62]. Another investigation Korenak et al. (2019) provides the feasibility of forward osmosis for dye wastewater treatment revealing 55% water recovery and greater than 94% COD rejection and 99% ionic rejection [77]. For congo red removal Khumalo et al. (2019) revealed that direct membrane distillation is also a pathway for removal of congo red examining the PTFE/PVDF thin membranes obtaining a flux of 0.0041 L/hm^2 [63]. However, membrane technology has a discrepant limitation of being highly cost-effective and energy-consuming. Therefore, the implications of such system are not preferred.

6.4 Treatment of Dye Emissions via Electrical Applications

There were electrical-driven processes like electrocoagulation, electrodialysis, capacitive deionization, etc. [78]. There were several conventional investigations using electrocoagulation of dye removal. But the exploration of capacitive deionization is wholesomely a new horizon to the field of wastewater treatment. So, the authors have attempted the same which has been delineated in preceding section.

7 Exploration of Capacitive Deionization for Dye Treatment: A Future Perspective

Capacitive deionization is an electrochemical process for the treatment of pollutants using two electrodes separated by an aqueous solution. The two terminals have been aligned in the system for positive and negative pole so as to adsorb the anionic and cationic species correspondingly [79]. Herein, there was simple rod-shaped electrode fabricated from activated carbon developed from the agricultural waste as synthesized by [1] using walnut shell activated carbon and bonded with polyvinylidene difluoride and using N-Methyl-2-pyrrolidone as solvent. The diameter of the electrode was kept around 0.5 cm and a copper wire (0.80 mm) was placed within the captive hole in

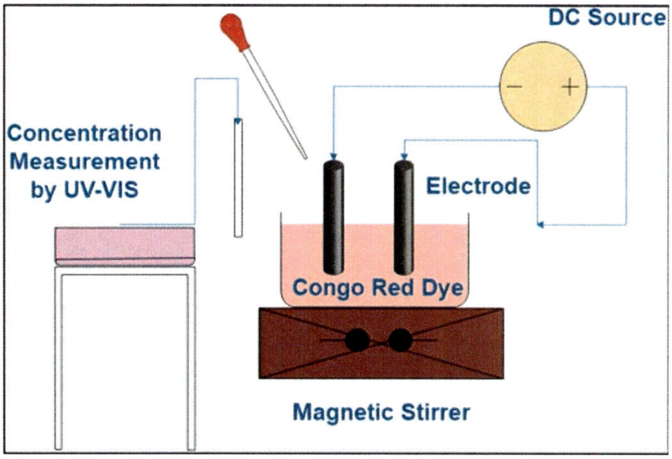

Fig. 5 Experimental setup for capacitive deionization of the congo red dye

between the electrode for passage of voltage. The connections were established, and alignment was built as shown in Fig. 5.

7.1 Outcomes and Discussions

The experiments were conducted for initial concentrations of 10, 20, 30, 40, 50 ppm congo red dye with a volume of 50 mL at room temperature. The application of 1.2 V was supplied to the system via DC source as aligned in Fig. 5. The % removal as obtained and the electrosorption capacities were evaluated as per below mathematical Eqs. 1 and 2 [66]:

$$\% \text{ Removal}, \ \%R = \frac{(C_{in} - C_{fi})}{C_{in}} \times 100 \tag{1}$$

$$\text{Electrosorption Capacity}, \ \ E_{cap} = \frac{(C_{in} - C_{fi})}{m_{el}} \times V_{ba} \tag{2}$$

where $\%R$, C_{in}, C_{fi}, E_{cap}, m_{el}, *and* V_{ba} are variables resembling percent removal (%), initial feed concentration (mg/L), final feed concentration (mg/L), electrosorption capacity (mg/g), electrode mass (g), the volume of batch (L), respectively.

The results obtained were as shown in Fig. 6. It was analysed that upon increasing the initial concentration the percentage removal was decreased [80]. The maximum percentage removal was obtained 41.8% at an initial concentration of 10 ppm after 2 h of contact period and the electrosorption capacity of 1.16 mg/g. Although there were high capacities reported in literature for congo red dye removal, but present work

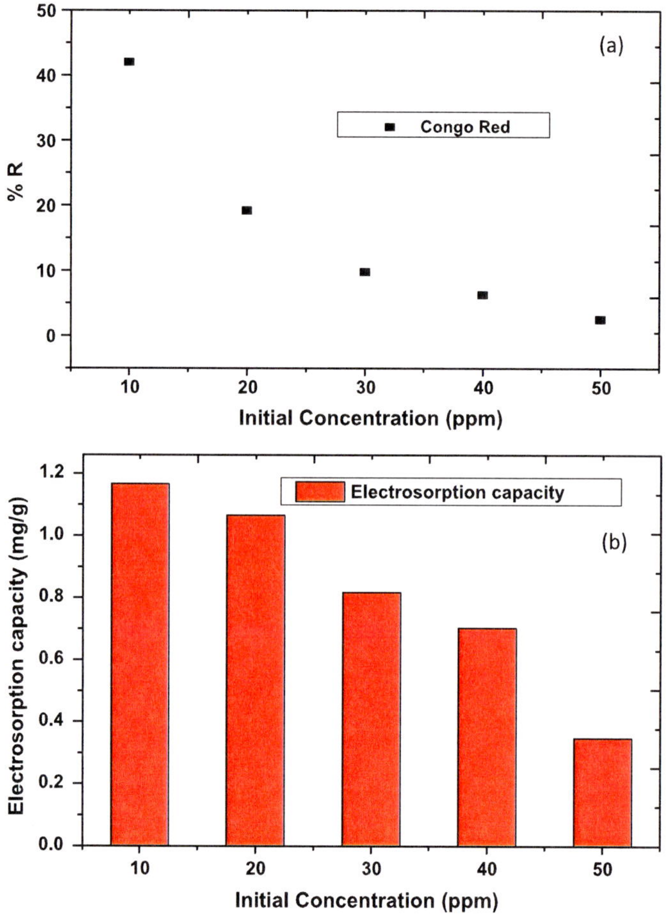

Fig. 6 **a** Percentage removal and **b** electro sorption capacities at varying initial concentration

has a nominal energy inclusion of 0.62 kWh m^{-3}. Moreover, further modifications in the material's surface may reveal better performance. The future work would be focused on the modifications and material improvements.

8 Conclusion

The present chapter reported the health hazards associated with the effluents discharged from the textile industries. Delineates the proliferation of dye used in textiles and presents the global issues and Indian scenario of discharges. It portrays the toxicological effects of dyes on microorganisms and humans. The article states the technologies for the treatment and proposes the new technology based on electrochemical treatment for the carcinogenic congo red dye which till date has not yet been attempted. Lastly, the performance was evaluated revealing 41.8% removal and 1.16 mg/g capacity at an initial concentration of 10 ppm. Concluding the article to the point that capacitive deionization could be a possible technology for the treatment of dye effluents.

References

1. Maheshwari K, Solanki YS, Ridoy MSH, Agarwal M, Dohare R, Gupta R (2020) Ultrasonic treatment of textile dye effluent utilizing microwave-assisted activated carbon. Environ Prog Sustain Energy. https://doi.org/10.1002/ep.13410
2. Agrawal M, Maheshwari K (n.d.) Textile water: characterization, environmental impact and treatment, vol 2, pp 47–75
3. Salleh MAM, Mahmoud DK, Karim WAWA, Idris A (2011) Cationic and anionic dye adsorption by agricultural solid wastes: a comprehensive review. Desalination 280:1–13. https://doi.org/10.1016/j.desal.2011.07.019
4. Li W, Mu B, Yang Y (2019) Feasibility of industrial-scale treatment of dye wastewater via bio-adsorption technology. Bioresour Technol 277:157–170. https://doi.org/10.1016/j.biortech.2019.01.002
5. Nasseh N, Arghavan FS, Rodriguez-Couto S, Hossein Panahi A (2020) Synthesis of FeNi3/SiO2/CuS magnetic nano-composite as a novel adsorbent for Congo Red dye removal. Int J Environ Anal Chem 7319. https://doi.org/10.1080/03067319.2020.1754810
6. Saini RD (2017) Textile organic dyes: polluting effects and elimination methods from textile waste water. Int J Chem Eng Res 9:975–6442 (2017). http://www.ripublication.com
7. Mary Ealias A, Saravanakumar MP (2019) A critical review on ultrasonic-assisted dye adsorption: mass transfer, half-life and half-capacity concentration approach with future industrial perspectives. Crit Rev Environ Sci Technol 49:1959–2015 (2019). https://doi.org/10.1080/10643389.2019.1601488
8. Bhat SA, Zafar F, Mondal AH, Mirza AU, Rizwanul Haq QM, Nishat N (2020) Efficient removal of Congo red dye from aqueous solution by adsorbent films of polyvinyl alcohol/melamine-formaldehyde composite and bactericidal effects. J Clean Prod 255:120062. https://doi.org/10.1016/j.jclepro.2020.120062
9. Al Prol AE (2019) Study of environmental concerns of dyes and recent textile effluents treatment technology: a review. Asian J Fish Aquat Res 3:1–18. https://doi.org/10.9734/ajfar/2019/v3i230032.
10. Routoula E, Patwardhan SV (2020) Degradation of anthraquinone dyes from effluents: a review focusing on enzymatic dye degradation with industrial potential. Environ Sci Technol 54:647–664. https://doi.org/10.1021/acs.est.9b03737
11. Hassaan MA, El Nemr A (2017) Health and environmental impacts of dyes: mini review. Am J Environ Sci Eng 1:64–67. https://doi.org/10.11648/j.ajese.20170103.11
12. Kumar A, Sehgal M (2018) Hydrogen fuel cell technology for a sustainable future: a review. SAE Tech Pap 2018:2588–2591. https://doi.org/10.4271/2018-01-1307
13. Azari A, Nabizadeh R, Nasseri S, Mahvi AH, Mesdaghinia AR (2020) Comprehensive systematic review and meta-analysis of dyes adsorption by carbon-based adsorbent materials: classification and analysis of last decade studies. Chemosphere 250:126238. https://doi.org/10.1016/j.chemosphere.2020.126238
14. Tkaczyk A, Mitrowska K, Posyniak A (2020) Synthetic organic dyes as contaminants of the aquatic environment and their implications for ecosystems: a review. Sci Total Environ 717:137222. https://doi.org/10.1016/j.scitotenv.2020.137222

15. Yeow PK, Wong SW, Hadibarata T (2021) Removal of Azo and Anthraquinone dye by plant biomass as adsorbent—a review. 11:8218–8232
16. Menon P, Singh TSA, Pani N, Nidheesh PV (2020) Chemosphere electro-fenton assisted sonication for removal of ammoniacal nitrogen and organic matter from dye intermediate industrial wastewater. Chemosphere 128739. https://doi.org/10.1016/j.chemosphere.2020.128739
17. Masarbo RS, Karegoudar TB (2020) Decolourisation of toxic azo dye Fast Red E by three bacterial strains: process optimisation and toxicity assessment. Int J Environ Anal Chem 00:1–11. https://doi.org/10.1080/03067319.2020.1759048
18. Qin Q, Ma J, Liu K (2009) Adsorption of anionic dyes on ammonium-functionalized MCM-41. 162:133–139. https://doi.org/10.1016/j.jhazmat.2008.05.016
19. Sarkar S, Ponce NT, Banerjee A, Bandopadhyay R, Rajendran S, Lichtfouse E (2020) Green polymeric nanomaterials for the photocatalytic degradation of dyes: a review. Environ Chem Lett. https://doi.org/10.1007/s10311-020-01021-w
20. Yazdani MR (2018) Engineered adsorptive materials for water remediation
21. Leaper S, Abdel-Karim A, Gad-Allah TA, Gorgojo P (2019) Air-gap membrane distillation as a one-step process for textile wastewater treatment. Chem Eng J 360:1330–1340. https://doi.org/10.1016/j.cej.2018.10.209
22. Gürses A, Açıkyıldız M, Güneş K, Gürses MS (2016) Dyes and pigments: their structure and properties. SpringerBriefs in green chemistry for sustainability, pp 13–29. https://doi.org/10.1007/978-3-319-33892-7
23. Mohebali S, Bastani D, Shayesteh H (2019) Equilibrium, kinetic and thermodynamic studies of a low-cost biosorbent for the removal of Congo red dye: acid and CTAB-acid modified celery (Apium graveolens). J Mol Struct 1176:181–193. https://doi.org/10.1016/j.molstruc.2018.08.068
24. Yaseen DA, Scholz M (2019) Textile dye wastewater characteristics and constituents of synthetic effluents: a critical review. Springer, Berlin, Heidelberg. https://doi.org/10.1007/s13762-018-2130-z
25. Benkhaya S, M' rabet S, El Harfi A (2020) A review on classifications, recent synthesis and applications of textile dyes. Inorg Chem Commun 115:107891. https://doi.org/10.1016/j.inoche.2020.107891
26. Tanvi DA, Lohit RT, Pratam KM, Vijayalakshmi BK, Devaraja TN, Mahanthesh V, Chakra PS, Devaraja G, Ramesh A (2020) Biosorption of heavy metal arsenic from Industrial Sewage of Davangere District, Karnataka, India, using indigenous fungal isolates. Appl Sci.https://doi.org/10.1007/s42452-020-03622-0
27. Eriksson E, Auffarth K, Eilersen AM, Henze M, Ledin A (2003) Household chemicals and personal care products as sources for xenobiotic organic compounds in grey wastewater. Water SA 29:135–146. https://doi.org/10.4314/wsa.v29i2.4848
28. Amin M, Khodabakhshi A (2012) Determination of malachite green in trout tissue and effluent water from fish farms. Int J Environ Health Eng 1:10. https://doi.org/10.4103/2277-9183.94394
29. li2013.pdf (n.d.)
30. Rahi RK, Prasad RN, Gupta V (2018) Analysis of physico-chemical properties of textile effluents collected from Sanganer, Jaipur. https://doi.org/10.36282/IJASRM/3.7.2018.627
31. Ahmed I, Haque F, Rahman MATMT, Parvez AK, Mou TJ (2020) Screening of Methyl Red degrading bacteria isolated from textile effluents of Savar Area, Dhaka, Bangladesh 301–318. https://doi.org/10.4236/abb.2020.117022
32. Khatun M, Mohammad G, Rahman MT, Kabir SE (2020) Organic waste removal from pharmaceutical and textile effluents using composite adsorbent 55:197–206. https://doi.org/10.3329/bjsir.v55i3.49393
33. Sani ZM, Dalhatu AS, Namadina MM, Abdullahi IL (2020) Impact assessment of dyeing processing activity on soils of selected sites in Kano Metropolis, Nigeria. 6:219–225
34. Zhou H, Zhou L, Ma K (2020) Science of the total environment micro fiber from textile dyeing and printing wastewater of a typical industrial park in China: occurrence, removal and release. Sci Total Environ 739:140329. https://doi.org/10.1016/j.scitotenv.2020.140329

35. Ahmad MT, Sushil M, Krishna M (2012) Influence of dye industrial effluent on physico chemical characteristics properties of soil at Bhairavgarh, Ujjain, MP, India. 1:50–53
36. Gupta K, Khatri OP (2019) Fast and efficient adsorptive removal of organic dyes and active pharmaceutical ingredient by microporous carbon: effect of molecular size and charge. Chem Eng J 378:122218. https://doi.org/10.1016/j.cej.2019.122218
37. Lokhande RS, Singare PU, Pimple DS (2011) Study on physico-chemical parameters of waste water effluents from Taloja Industrial Area of Mumbai, India. 1:1–9. https://doi.org/10.5923/j.ije.20110101.01
38. Ranganathan K, Karunagaran K, Sharma DC (2007) Recycling of wastewaters of textile dyeing industries using advanced treatment technology and cost analysis—case studies. 50:306–318. https://doi.org/10.1016/j.resconrec.2006.06.004
39. Dubey SK, Yadav R, Chaturvedi RK, Yadav RK, Sharma VK, Minhas PS (2010) Contamination of ground water as a consequence of land disposal of dye waste mixed sewage effluents : a case study of Panipat District of Haryana, India. 295–300. https://doi.org/10.1007/s00128-010-0073-2
40. Kumar S, Walia YK (2018) Study on physico-chemical parameters of water from different dyeing units effluents in Ludhiana City, India 4:145–150
41. Gupta R, Jindal T, Khan AS, Srivastava P, Kanaujia A (2019) Assessment of ground water quality and its suitability for drinking in industrial area Jajmau, Kanpur, India. 19:1569–1571
42. Mostafa MRIMG (2020) Characterization of textile dyeing effluent and its treatment using polyaluminum chloride. Appl Water Sci 10:1–10. https://doi.org/10.1007/s13201-020-01204-4
43. Madamwar D, Onkar T, Jain K (2019) Mapping of research outcome on remediation of dyes, dye intermediates and textile industrial waste. 1–228. http://dbtindia.gov.in/sites/default/files/Textile_Dyes_Compendium-Final.pdf
44. Dey S, Islam A (2015) A review on textile wastewater characterization in Bangladesh. Academia.edu. Resour Environ 5:15–44. https://doi.org/10.5923/j.re.20150501.03
45. Samir S, Al-tohamy R, Koutra E, El-naggar AH, Kornaros M, Sun J (2021) Valorizing lignin-like dyes and textile dyeing wastewater by a newly constructed lipid-producing and lignin modifying oleaginous yeast consortium valued for biodiesel and bioremediation. J Hazard Mater 403:123575. https://doi.org/10.1016/j.jhazmat.2020.123575
46. Methneni N, Anthonissen R, Van De Maele J, Trifa F, Verschaeve L, Ben Mansour H (2020) Assessment of natural coagulants to remediate Tunisian textile wastewater by combining physicochemical, analytical, and toxicological data
47. Li F, Huang J, Xia Q, Lou M, Yang B, Tian Q, Liu Y (2018) Separation and purification technology direct contact membrane distillation for the treatment of industrial dyeing wastewater and characteristic pollutants. Sep Purif Technol 195:83–91. https://doi.org/10.1016/j.seppur.2017.11.058
48. Rathore J (2012) Studies on pollution load induced by dyeing and printing units in River. Int J Environ Sci 3:735–742. https://doi.org/10.6088/ijes.2012030131072
49. Sharma N, Sharma SK, Gehlot A (2014) Influence of dyeing and printing industrial effluent on physicochemical Characteristics of water—case study on the printing cluster of Bagru, Jaipur (Rajasthan), India. IOSR J Appl Chem 7:61–64. https://doi.org/10.9790/5736-07416164
50. Reddy S, Osborne WJ (2020) Heavy metal determination and aquatic toxicity evaluation of textile dyes and effluents using Artemia salina. Biocatal Agric Biotechnol 25:101574. https://doi.org/10.1016/j.bcab.2020.101574
51. Novotný Č, Dias N, Kapanen A, Malachová K, Vándrovcová M, Itävaara M, Lima N (2006) Comparative use of bacterial, algal and protozoan tests to study toxicity of azo- and anthraquinone dyes. Chemosphere 63:1436–1442. https://doi.org/10.1016/j.chemosphere.2005.10.002
52. Gottlieb A, Shaw C, Smith A, Wheatley A, Forsythe S (2003) The toxicity of textile reactive azo dyes after hydrolysis and decolourisation. J Biotechnol 101:49–56. https://doi.org/10.1016/S0168-1656(02)00302-4
53. Vinitnantharat S, Chartthe W, Pinisakul A (2008) Toxicity of reactive red 141 and basic red 14 to algae and waterfleas. Water Sci Technol 58:1193–1198. https://doi.org/10.2166/wst.2008.476

54. de Luna LAV, da Silva THG, Nogueira RFP, Kummrow F, Umbuzeiro GA (2014) Aquatic toxicity of dyes before and after photo-Fenton treatment. J Hazard Mater 276:332–338. https://doi.org/10.1016/j.jhazmat.2014.05.047
55. Croce R, Cinà F, Lombardo A, Crispeyn G, Cappelli CI, Vian M, Maiorana S, Benfenati E, Baderna D (2017) Aquatic toxicity of several textile dye formulations: acute and chronic assays with Daphnia magna and Raphidocelis subcapitata. Ecotoxicol Environ Saf 144:79–87. https://doi.org/10.1016/j.ecoenv.2017.05.046
56. Aksu O, Yildirim NC, Danabas D, Yildirim N (2017) Biochemical impacts of the textile dyes Remazol Brillant Blue R and Congo Red on the crayfish Astacus leptodactylus (Decapoda, Astacidae). Crustaceana 90:1563–1574. https://doi.org/10.1163/15685403-00003738
57. Chung K-T (2016) Azo dyes and human health: a review. Environ Sci Health Care 34:1–60. https://doi.org/10.1080/10590501.2016.1236602
58. Feng J, Cerniglia CE, Chen H (2012) Toxicological significance of azo dye metabolism by human intestinal microbiota. Front Biosci Elit 4:568–586. https://doi.org/10.2741/e400
59. Chequer FMD, Angeli JPF, Ferraz ERA, Tsuboy MS, Marcarini JC, Mantovani MS, de Oliveira DP (2009) The azo dyes Disperse Red 1 and Disperse Orange 1 increase the micronuclei frequencies in human lymphocytes and in HepG2 cells. Mutat Res Genet Toxicol Environ Mutagen 676:83–86. https://doi.org/10.1016/j.mrgentox.2009.04.004
60. Sathishkumar K, AlSalhi MS, Sanganyado E, Devanesan S, Arulprakash A, Rajasekar A (2019) Sequential electrochemical oxidation and bio-treatment of the azo dye congo red and textile effluent. J Photochem Photobiol B Biol 200:111655. https://doi.org/10.1016/j.jphotobiol.2019.111655
61. Lee YH, Jeong JY, Jegal J, Mo JH (2008) Preparation and characterization of polymer-carbon composite membranes for the removal of the dissolved salts from dye wastewater. Dye Pigment 76:372–378. https://doi.org/10.1016/j.dyepig.2006.08.049
62. Baghel R, Upadhyaya S, Chaurasia SP, Singh K, Kalla S (2018) Optimization of process variables by the application of response surface methodology for naphthol blue black dye removal in vacuum membrane distillation. J Clean Prod 199:900–915. https://doi.org/10.1016/j.jclepro.2018.07.214
63. Khumalo NP, Nthunya LN, De Canck E, Derese S, Verliefde AR, Kuvarega AT (2019) Separation and purification technology Congo red dye removal by direct membrane distillation using PVDF/PTFE membrane. Sep Purif Technol 211:578–586. https://doi.org/10.1016/j.seppur.2018.10.039
64. Heidari MR, Varma RS, Ahmadian M, Pourkhosravani M, Asadzadeh SN, Karimi P, Khatami M (2019) Photo-Fenton like catalyst system: activated carbon/CoFe 2 O 4 nanocomposite for reactive dye removal from textile wastewater. Appl Sci 9. https://doi.org/10.3390/app9050963
65. Mook WT, Aroua MK, Szlachta M, Lee CS (2017) Optimisation of Reactive Black 5 dye removal by electrocoagulation process using response surface methodology. Water Sci Technol 75:952–962. https://doi.org/10.2166/wst.2016.563
66. Chen W, Yan X (2020) Progress in achieving high-performance piezoresistive and capacitive flexible pressure sensors: a review [J]. J Mater Sci Technol 43(0):175–188
67. Zhou Y, Lu J, Zhou Y, Liu Y (2019) Recent advances for dyes removal using novel adsorbents: a review. Environ Pollut 252:352–365. https://doi.org/10.1016/j.envpol.2019.05.072
68. Silva TL, Cazetta AL, Souza PSC, Zhang T, Asefa T, Almeida VC (2018) Mesoporous activated carbon fibers synthesized from denim fabric waste: efficient adsorbents for removal of textile dye from aqueous solutions. J Clean Prod 171:482–490. https://doi.org/10.1016/j.jclepro.2017.10.034
69. Mahmoodi NM, Masrouri O, Arabi AM (2014) Synthesis of porous adsorbent using microwave assisted combustion method and dye removal. J Alloys Compd 602:210–220. https://doi.org/10.1016/j.jallcom.2014.02.155
70. Borousan F, Youse F, Ghaedi M (2019) Removal of Malachite Green Dye using IRMOF-3–MWCNT-OH–Pd-NPs as a novel adsorbent: kinetic, isotherm, and thermodynamic studies (2019). https://doi.org/10.1021/acs.jced.9b00298

71. Liu Q, Gao Y, Zhou Y, Tian N, Liang G, Ma N, Dai W (2019) Highly improved water resistance and Congo red uptake capacity with a Zn/Cu-BTC@MC composite adsorbent. J Chem Eng Data. https://doi.org/10.1021/acs.jced.9b00159
72. Zolgharnein J, Dermanaki Farahani S, Bagtash M, Amani S (2020) Application of a new metal-organic framework of [Ni2F2(4,4′-bipy)2(H2O)2](VO3)2.8H2O as an efficient adsorbent for removal of Congo red dye using experimental design optimization. Environ Res 182:109054. https://doi.org/10.1016/j.envres.2019.109054.
73. Dalvand A, Gholibegloo E, Ganjali MR, Golchinpoor N, Khazaei M, Kamani H, Hosseini SS, Mahvi AH (2016) Comparison of Moringa stenopetala seed extract as a clean coagulant with Alum and Moringa stenopetala-Alum hybrid coagulant to remove direct dye from Textile Wastewater. Environ Sci Pollut Res 23:16396–16405. https://doi.org/10.1007/s11356-016-6708-z
74. Shankar YS, Ankur K, Bhushan P (n.d.) Utilization of water treatment plant (WTP) sludge for pretreatment of dye wastewater using coagulation/Flocculation. Springer, Singapore. https://doi.org/10.1007/978-981-13-0215-2
75. Le OTH, Tran LN, Doan VT, Van Pham Q, Van Ngo A, Nguyen HH (2020) Mucilage extracted from dragon fruit peel (Hylocereus undatus) as flocculant for treatment of dye wastewater by coagulation and flocculation process. Int J Polym Sci. https://doi.org/10.1155/2020/7468343
76. Nataraj SK, Hosamani KM, Aminabhavi TM (2009) Nano filtration and reverse osmosis thin film composite membrane module for the removal of dye and salts from the simulated mixtures. Desalination 249:12–17. https://doi.org/10.1016/j.desal.2009.06.008
77. Korenak J, Hélix-Nielsen C, Bukšek H, Petrinić I (2019) Efficiency and economic feasibility of forward osmosis in textile wastewater treatment. J Clean Prod 210:1483–1495. https://doi.org/10.1016/j.jclepro.2018.11.130
78. Moussa DT, El-Naas MH, Nasser M, Al-Marri MJ (2017) A comprehensive review of electrocoagulation for water treatment: potentials and challenges. J Environ Manag 186:24–41. https://doi.org/10.1016/j.jenvman.2016.10.032
79. Baroud TN, Giannelis EP (2018) High salt capacity and high removal rate capacitive deionization enabled by hierarchical porous carbons. Carbon NY 139:614–625. https://doi.org/10.1016/j.carbon.2018.05.053
80. Xu X, Tan H, Wang Z, Wang C, Pan L, Kaneti YV, Yang T, Yamauchi Y (2019) Extraordinary capacitive deionization performance of highly-ordered mesoporous carbon nano-polyhedra for brackish water desalination. Environ Sci Nano 6:981–989. https://doi.org/10.1039/c9en00017h

The Utilization of Biomaterials for Water Purification: Dyes, Heavy Metals, and Pharmaceuticals

Ali Khadir, Mehrdad Negarestani, Asiyeh kheradmand, Armin Azad, and Mika Sillanpää

Abstract Industrialization, modernization and improper agricultural practices have led to the discharge of many diverse pollutants in the environment, threatening the lives of humans and the ecosystem. Water pollution is one of the most important issues that humans are dealing with. Up to now, many attempts have been conducted to remediate and treat polluted water. Environmentally biomaterials are of great alternative for water purification, particularly due to their promising properties, including cost-effectiveness, availability, non-toxicity, and excellent performance. Accordingly, the present chapter explores the utilization of some biomaterials, namely, chitosan, Luffa, sisal, and hydroxyapatite within the field of water and wastewater technology for the removal of dyes, heavy metals, and pharmaceuticals. The consideration of these three groups of pollutants is mainly because of the fact that in actual water media, these may come all together. In addition, the fundamentals of the adsorption process as well as isotherms/kinetics models and thermodynamic parameters are well discussed.

Keywords Water purification · Biomaterial · Chitosan · Luffa · Sisal · Hydroxyapatite

A. Khadir (✉)
Young Researcher and Elite Club, Yadegar-e-Imam Khomeini (RAH) Shahre Rey Branch, Islamic Azad University, Tehran, Iran

M. Negarestani · A. kheradmand
Department of Civil and Environmental Engineering, Iran University of Science and Technology, Tehran, Iran

A. Azad
Department of Water Engineering and Hydraulic Structures, Faculty of Civil Engineering, Semnan University, 35131-19111 Semnan, Iran

M. Sillanpää
Department of Biological and Chemical Engineering, Aarhus University, Nørrebrogade 44, 8000 Aarhus C, Denmark

Institute of Research and Development, Duy Tan University, Da Nang 550000, Vietnam

© The Author(s), under exclusive license to Springer Nature Singapore Pte Ltd. 2021

S. S. Muthu and A. Khadir (eds.), *Novel Materials for Dye-containing Wastewater Treatment*, Sustainable Textiles: Production, Processing, Manufacturing & Chemistry,
https://doi.org/10.1007/978-981-16-2892-4_2

1 Introduction

With the advent of technology in every aspect of life, humans decided to move toward industrialization and a life of high quality and standard. Even a great flourish has observed in agriculture, leading to more efficient productivity. Such advancement and progress are not always without any disadvantages. Human activities have adversely affected the ecosystem and living environment by imposing a great deal of pollutants, releasing them freely into water, soil, and air. 'Water pollution', 'Soil pollution', and 'Air pollution' are words that every individual has heard for ages. Water pollution defines as the entrance of any harmful substances (generally chemicals and microorganisms) to water media, leading to lowering its quality and posing hazards for humans or the environment. By contamination water media, some consequences are occurred in water, including temperature and color change, undesired acidity or alkalinity, unpleasant odor, turbidity by organic/inorganic solids, etc. Although natural sources contaminate the environment and also water, human activities and sources are the primary reasons for the current concern for water pollution.

Many scholars have already warned the nations about the adverse effects of pollutants in the environment. Psychic disturbances, cardiovascular effects, Landry's paralysis, Minamata disease, kidney damage, blood vessel congestion, potent neurotoxicity, anemia, brain tissue damage, bone degeneration, renal dysfunction, etc. are some of the effects of metals, including thallium, mercury, cadmium, and lead [1]. In case of dyes and textile industry effluent, their wastewaters frequently own a high amount of biochemical oxygen demand (BOD) and chemical oxygen demand (COD) that are hard to degrade. Also, dye-containing wastewaters tend to accumulate on the surface of the water, preventing light penetration and a reduction in the rate of photosynthesis and dissolved oxygen [2]. Pharmaceuticals, particularly antibiotics could pollute the ecosystem by propagating antibiotic resistance bacteria and antibiotic resistance genes propagation [3, 4].

Considering the mentioned hazards for ecosystem and living creatures, treatment of contaminated water is of high priority. During the last decades, many nations have attempted to mitigate the negative effects of water pollution by utilizing different water purification techniques. Frequently, water and wastewater treatment techniques fall into physical, chemical, and biological approaches in which a combination of these is required to reach a sustainable removal and efficiency. Adsorption [5], electrocoagulation [6, 7], photocatalytic degradation [4, 8], membrane separation [9], aerobic/anaerobic biodegradation [10], etc. Adsorption is a physical mass transfer phenomenon that has gained a great deal of attention during the last two decades, and many believed that this technique could be an excellent alternative for pollutant elimination [11, 12]. Adsorption provides a few particular advantages: simplicity, high removal rate, cost-effectiveness, simplicity in operation, and reusability [3]. Adsorbents employ various mechanisms to remove the pollutants such as acid–base interaction, chelation, complexation, hydrogen bonding, electrostatic interactions, diffusion, ion-exchange, hole filling, and π interactions. Note that the appropriateness and favorability of the adsorption process may be enhanced by using low-cost materials as the adsorbent. Khadir et al., for instance, suggested Luffa composite as a biomaterial for the treatment of water contaminated with ibuprofen molecules [13].

The utilization of biomaterial has been a fascinating and interesting option for many studies since it provides affordable, effective, and bio-removal separation techniques. Bioadsorbents or biosorbents are in fact biomaterials that are derived from biological sources and could exhibit the adsorption capacities at par with chemical adsorbents [14]. Chief bio-macromolecules, including cellulose, hemicellulose, proteins, and lignin, are the main constituents of the biomaterials. The presence of carboxyl, carbonyl, hydroxyl, amine and amide as the reactive functional groups enables bio-adsorbents to establish the required bond with the pollutants [15]. According to the present literature, bioadsorbents have been applied for the removal of heavy metals, dyes, pesticides, and pharmaceuticals due to their inexpensiveness, abundancy and easy to find, various functional groups, structural heterogeneity, pore sizes, low operating costs, minimal volumes of chemical disposal, etc. [14, 16].

The current chapter aims at the following goals:

- A fundamental introduction to the adsorption process and its mechanism.
- Adsorption isotherms (Langmuir/Freundlich assumptions) for sorption systems.
- Kinetics models (Pseudo-first order, Pseudo-second order, and Elovich) for sorption systems.
- Thermodynamic parameters and their importance.
- A comprehensive discussion on some of the most famous and applicable biomaterials and biosorbent in the field of water depollution.

2 Fundamentals of the Adsorption Process

Since the utilization of the first water and wastewater treatment techniques, such as sedimentation and simple filtration, many people and companies have tried to introduce better and more efficient methods for the purification of contaminated waters. Aside from the fact that more energy savings' systems along with excellent performance are desired, it is urgent to note that some of the conventional methods of treatment are not good enough to remove a wide range of pollutants, especially emerging ones, including pharmaceuticals, or somehow dyes. Many reports have already been released, demonstrating that present/conventional water and wastewater treatment plants are not capable of removing all emerging pollutants from the influent [3, 17, 18]. Hence, these hazardous substances enter the environment via effluent from the treatment plant, contaminating the ecosystem. Pharmaceuticals, for instance, have been detected in drinking water and groundwater. This has led to the utilization of novel and advanced systems for the purification of water and wastewater.

The adsorption process has been nominated as a promising, efficient, simple, and cost-effective water purification technology that has been extensively utilized during the past decade to remove various kinds of pollutants. Adsorption, in fact, is a physical transport of the pollutant molecules (known as adsorbate) from the liquid phase to a solid, porous or functionalized surface (known as adsorbent) [19]. In 1550 BC,

Egyptians employed charcoal for medicinal purposes, and it was actually of the first utilization of the adsorption process. This separation technique gained significant attention once zeolite was synthesized commercially in 1956 [20]. However, it is about twenty years that many scholars are attempting to improve the performance of the sorption system by introducing more engineered materials.

The interaction of the adsorbate and adsorbent is the key factor affecting the efficiency of the entirety of the process. Adsorption is divided into physical adsorption and chemical adsorption. If the adsorbate is adsorbed via London–van der Waals forces, it is called physical sorption or physisorption. In this case, the attractive forces are weak and usually the process is exothermic in nature. Physisorption is a weak force and acts non-selectively, leading to the formation of multi-layers on the adsorbent surface. If chemical bonds controlled the attachment of the adsorbate to the adsorbent, the process is chemical adsorption or chemisorption [21]. This type of adsorption tends to form a mono-layer on the adsorbent. Also, it is energetic and irreversible.

There are multiple operational parameters that affect the outcome of the adsorption process. Contact time is a primary factor which means enough time must be given to the process until the equilibrium state is reached. In addition, by increasing the contact time between adsorbent and adsorbate, it is believed that adsorption efficiency might increase with varying rates, depending on the available sites of the material [5]. Adsorbent dosage, the amount of adsorbent in the solution, is also critical in the sorption system since it influences the efficiency and sludge production. A higher dosage of adsorbent could provide a greater number of empty sites that adsorbate could occupy, leading to an increment in the adsorption efficiency. It is noted that agglomeration and aggregation of the adsorbent particles, especially nano-materials, can reduce the surface area of the adsorbent which is not beneficial in terms of the adsorption process. Next is the initial pollutant concentration. If pollutant concentration increases, it is expected that the efficiency of the process would decline under constant adsorbent dosage. It is tried to keep the pollutant concentration to a minimal value. Salinity, temperature, competing ions/compounds, etc. are other parameters determining the efficiency of the adsorption system. In order to reach the maximum capacity of the treatment, these parameters must be optimized.

3 Adsorption Isotherms

Approximately a century ago, equations called isotherms were suggested to expound the process of the sorption system. These models are based on the equilibrium relationship between the amount of pollutant adsorbed and its concentration in the solution under constant temperature. Equilibrium experiments are those tests that at a certain time the number of molecules entering the surface of the adsorbent are equal to the number of the molecules that are leaving the adsorbent surface. This behavior occurs once the adsorbent is saturated, meaning that no more pollutants cannot be removed anymore. Isotherm models are applied to saturated systems. Langmuir and Freundlich are the two most important models in adsorption studies.

Langmuir model is one of the simplest equations of isotherm, assuming that the surface possesses a certain number of binding sites with equal energy/heat and no lateral interactions. Monolayer formation is desired by Langmuir isotherm. For a pure component system and multicomponent mixtures systems, the Langmuir equation might be written as follows [20]:

$$q_e = \frac{q_s K C}{1 + K C},$$

$$q_e = \frac{q_{m,\ j} K_j C_j}{1 + \sum_{i=1}^{i=n} K_i C_i},$$

where q_e (mg/g) and q_s (mg/g) are the equilibrium adsorption capacity and Langmuir maximum adsorption capacity, respectively. K is the Langmuir constant. The index j is indicative of the presence of 'n' species in the solution. With the contribution of Langmuir, a dimensionless equilibrium parameter (R_L) could be calculated as follow:

$$R_L = \frac{1}{1 + K_L C_0}.$$

According to the value of R_L, four different possibilities are feasible:

$$0 < R_L < 1 \rightarrow \text{favorable},$$
$$R_L > 1 \rightarrow \text{unfavorable},$$
$$R_L = 1 \rightarrow \text{linear},$$
$$R_L = 0 \rightarrow \text{irreversible}.$$

Freundlich isotherm is an empirical adsorption model that has been extensively employed in environmental investigations. This model is based on the formation of multilayers over a heterogeneous surface with the different free energy of sorption. Freundlich isotherm is expressed as follows [6]:

$$q_e = K_F C_e^{\frac{1}{n}},$$

where K_F is the Freundlich constant, which represents adsorption capacity; $1/n$ is an empirical constant indicating the adsorption intensity of the system. If $n = 1$, then adsorption is linear; if $n < 1$, then adsorption is a chemical process; if $n > 1$, then adsorption is a physical process [22].

4 Adsorption Kinetic Models

To figure out more information about the adsorption process and involved mechanism, accurate kinetic modeling is necessary in order to draw credible conclusions. So, the effect of contact time is one of the most important tests during optimization procedure that requires great care to obtain the best results. Generally speaking, these models contribute to finding the rate of the reactions, the amount of time required to attain the desired performance, and the scale of an adsorption apparatus [23].

In an adsorption process, different phenomena occur with different rates that the slowest rate determines the limiting rate. Once the adsorbate is exposed to the adsorbent, three steps may happen during the removal of the pollutant via sorption: (I) movement of the adsorbate from the bulk solution to the external surface of the adsorbent (external mass transfer), (II) diffusion of the adsorbate to the sites (internal diffusion), and (II) occurrence of sorption [24]. Over many years, a great number of kinetics models have been formulated for different sorption systems that consider the mentioned step of the adsorption. Among these models, Pseudo-first order and Pseudo-second order are the most common ones.

Lagergren (1898), the one who proposed the earliest model of kinetic in the adsorption system, investigated the adsorption of oxalic acid and malonic acid onto the charcoal and expressed the following equation:

$$\frac{dq_t}{dt} = K_1(q_e - q_t).$$

The non-linear and linear forms can be written as follows:

$$q_t = q_e\left(1 - e^{-K_1 t}\right),$$
$$\log(q_e - q_t) = -\frac{K_1}{2.303}t + \log q_e,$$

where K_1 (1/min) **is** the pseudo-first-order rate constant for the kinetic model.

The importance of Pseudo-second order cannot be ignored since so many investigations have believed that this model is superior to Pseudo-first order or Elovich describing the sorption process. Only some special studies declared that Pseudo-second order is not the primary model for the system [25]. Pseudo-second order model has the following form [26]:

$$\frac{dq_t}{dt} = K_2(q_e - q_t)^2.$$

The non-linear and linear forms are as follows:

$$\frac{t}{q_t} = \left(\frac{1}{q_e}\right)t + \frac{1}{K_2 q_e^2},$$

$$\frac{1}{q_t} = \left(\frac{1}{q_e}\right) + \left(\frac{1}{K_2 q_e^2}\right)\frac{1}{t},$$

$$q_t = -\left(\frac{1}{K_2 q_e}\right)\frac{q_t}{t} + q_e,$$

$$\frac{q_t}{t} = -(K_2 q_e) q_t + K_2 q_e^2.$$

Elovich is another second-order kinetic model for heterogeneous chemisorption of gases on solid surfaces. Accurate data at short reaction times are essential for the Elovich model [27]. The Elovich equation is generally expressed as follows [28]:

$$q_t = \frac{1}{\beta}\ln(\alpha\beta) + \frac{1}{\beta} lnt,$$

where α is the initial sorption rate of the Elovich equation (g/g min), and β is the adsorption constant (g/g).

5 Adsorption Thermodynamics

Once the thermodynamic study is performed, the adsorbent and adsorbate are interacted with each other under different temperatures while the other parameters are kept constant. The obtained results could provide substantial information about the interactions between the involved species. Thermodynamic parameters (free energy change ($\Delta G°$), entropy ($\Delta S°$) and enthalpy ($\Delta H°$)) are calculated by the effect of temperature on the adsorption process. Enthalpy declares the strength of the adsorption, and entropy indicates the change in the overall order of the system [29]. According to the laws of thermodynamics, the following equation could be expresses [30]:

$$\Delta G^\circ = \Delta H^\circ - T\Delta S^\circ,$$

$$\Delta G^\circ = -RT \ln K_{eq}.$$

By combining the last two equations, the well-known van't Hoff equation is found as follows:

$$\ln K_{eq} = -\frac{\Delta H^\circ}{RT} + \frac{\Delta S^\circ}{R}.$$

$\Delta S°$ and $\Delta H°$ are calculated from the slope and intercept of the van't Hoff equation. The equilibrium constant (K_{eq}) can be determined with the help of isotherm constants (Langmuir, Freundlich, Henry, or Flory–Huggins), partition constant, and distribution constant [26]. Since K_{eq} is dimensionless, many scholars have proposed different methods of changing other dimension-based constants to dimensionless ones. Langmuir isotherm, for instance, owns the constant K_L with the units of L/mg or L/mmol which could not directly be employed in the thermodynamics equation. The following formula could relate K_L with K_{eq} [26]:

$$K_{eq} = 55.5 \times 1000 \times K_L \left(\frac{\text{L}}{\text{mmol}}\right),$$

$$K_{eq} = 10^6 K_L \left(\frac{\text{L}}{\text{mg}}\right).$$

The positive values of ΔH° indicate the endothermic nature of the reaction, meaning that adsorption efficiency increases by increasing the temperature. However, exothermic reactions and more efficiency at a lower temperature are derived from negative values of ΔH°. In addition, as the ΔH° values increase the bonds between adsorbate and adsorbent tend to be chemisorption; however, low ΔH° values exhibit weak physical adsorption. The positive ΔS° demonstrates an increase in disorder at the interface during adsorption. The negative and positive values of ΔG° reveal spontaneous (thermodynamically favorable) and non-spontaneous adsorption.

6 Biomaterials and Biocompounds for Water Purification

6.1 *Chitosan*

Chitin and chitosan (Fig. 1) are two famous biomaterials, having extensive applications. The discovery of chitin bio-polymer was made from mushrooms in 1811 by Henri Braconnot, a French professor of natural history. Chitin was also observed in

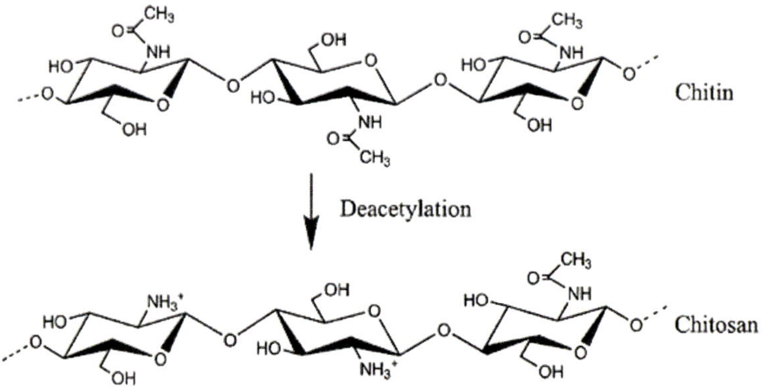

Fig. 1 Structure of chitin and chitosan

insects in 1823 by Odier. Chitosan, the deacetylated product of chitin, was synthesized in 1859 by Professor C. Rouget, and research on chitosan was intensified in the early twentieth century [31]. Chitosan is a versatile polysaccharide widely distributed in nature (second most abundant biopolymers after cellulose) produced by alkaline N-deacetylation of chitin. Chitosan is known as a biomaterial due to its biocompatibility, biodegradability, and non-toxic properties. Many application fields are described in scientific publications regarding the use of chitin, chitosan and their derivatives. Wastewater treatment and water using chitin or chitosan is an important application. Today the development of chitosan-based nanomaterials and nanocomposites is of great scope to exploit the unique properties of nano-adsorbents for water purification purposes.

Chitosan is a type of copolymer which consists of two repeating units, i.e., N-acetyl-2-amino-2-d-glucopyranose and 2-amino-2-deoxy-d-glucopyranose [32, 33]. These two repeating units are linked by the β-(1 → 4)-glycosidic bond. Because of this type of arrangement, chitosan shows a rigid crystalline structure through inter- and intra-molecular hydrogen bonding. Chitosan has one $-NH_2$ group and two –OH groups on each and every glucosidic residue. Owing to the presence of these two reactive groups, chitosan exhibits marvelous chemical as well as biological properties. The intrinsic pK_a value of chitosan strictly depends upon the degree of deacetylation, ionic strength, and the charge neutralization of $-NH_2$ groups. Practically, for fully neutralized amine functions with a degree of acetylation of not more than 50%, the pK_a value always lies between 6.3 and 6.7. The majority of naturally occurring polysaccharides such as pectin, dextrin, agar, agarose, carrageenan, and cellulose are acidic in nature, while chitosan is a highly basic polysaccharide. Chitosan exhibits special properties such as viscosity, solubility in various media, polyelectrolyte behavior, ability to form films, metal chelations, optical, and structural characteristics.

Many scholars have noted the fact that chitin and chitosan could be great in removing pollutants from polluted water and wastewater. They could be used as coagulating/flocculating agents and adsorbents for the removal of distinctive organic/inorganic matters. For instance, Cu^{+2} removal was investigated by different composites of chitosan, including chitosan beads, chitosan/polyaniline beads, chitosan/glutaraldehyde beads, and chitosan/alginate beads, owning the adsorption capacities of 52.6, 100, 31.20, and 67.66 mg/g, respectively [34]. Jiang et al. tested the performance of crosslinked glucan/chitosan for the removal of heavy metals [35]. The composite had the structure of pale brown-yellow honey-comb morphology. The authors pointed out heavy metal adsorption initially increased very rapidly in 40 min, and the saturation point was reached at 60 min. The effect of pH was also studied from one to seven, resulting in an increase in heavy metal adsorption. The elevated temperature was beneficial for the sorption performance, resulting in an endothermic process. The adsorption capacities of glucan/chitosan hydrogels for Cu^{+2}, Co^{+2}, Ni^{+2}, Pb^{+2} and Cd^{+2} were 342, 232, 184, 395 and 269 mg/g, respectively. They noted that the whole removal procedure was spontaneous and single-layer chemisorption.

In another investigation, the effect of ball milling on dye adsorption by chitosan (bought from Sigma-Aldrich, Germany) was studied by Qiu et al. [36]. When chitosan was not ball milled, its surface area was 0.922 m^2/g; however, after 0.5, 1, 2, and 3 h milling, the surface area of the chitosan was 2.038, 2.431, 2.073, and 1.803 m^2/g, respectively. It is considered that during the first hour of milling, the particles are breaking without aggregation; however, after an hour, the number of particles reduced, and agglomeration may occur. The utilization of 1 h ball milled chitosan, resulted in an adsorption capacity of 1.54 mmol/g for azo-dye reactive red 2 which was 70% higher than that of unmilled chitosan. For other dyes, namely, brilliant blue, sunset yellow and tartrazine, chitosan proved promising adsorption capacities in the range of 406.19 to 814.27 mg/g, 924.88 to 1432.98 mg/g, and 611.27 to 1065.55 mg/g, respectively [37].

M. N. dos Santos et al. prepared $ZnFe_2O_4$/chitosan magnetic particles (57.4 m^2/g) for the removal of diclofenac from water and found that adsorption capacities and removal percentages increased as initial pH values decreased [38]. This behavior was attributed to acidic behavior of diclofenac ($pK_a = 4$) and the protonated amino groups of chitosan ($NH_2 \rightarrow NH_3^+$). Similar observation between diclofenac and chitosan-based material was reported in the literature [39]. The low dosage of $ZnFe_2O_4$/chitosan resulted in high adsorption capacity; at a dosage of 0.2 g/L, the adsorption capacity was 180 mg/g. The kinetic data were analyzed by Pseudo-first and Pseudo-second order models, and based on R^2 and average relative error values, the Pseudo-second order model was the best one. High adsorption capacity and adsorbent reusability were two primary advantages of the synthesized material. In case of Bisphenol A removal, the removal efficiency of 93.8% was reported by means of chitosan-based composite [40]. Table 1 summarized the adsorption capacities of chitosan toward dyes, pharmaceuticals, and heavy metals.

6.2 Luffa

Luffa is a plant with pale to bright yellow flowers belonging to the family *Cucurbitaceae* and generally grows in tropical and subtropical areas around the world [61]. *L. cylindrica* and *L. acutangula* are two primary species of Luffa. In many regions such as Iran, Korea, China, Japan, India, Brazil, Australia, Africa, and America, this plant could be found. China is considered to be one of the leading countries in Luffa plantation, particularly in areas such as Jiangxi, Henan, Sichuan, Guangdong, Jiangsu, and Anhui [62]. In Japan also Luffa was used for the manufacture of filters in 1890–1895. Regarding different environmental conditions, this plant has different sizes ranging from cm-scale to even 9 m [63]. The diameter and density are 8–30 μm and 1.48 g/cm^3, respectively. Figure 2 shows the Luffa. Cellulose (60%), hemicellulose (30%), and lignin (10%) are the general component of Luffa that could fluctuate in the range of 50–60% for cellulose, 25–28% for hemicellulose and 10–12% for lignin. In regions with suitable light, lignin could have better synthesis, resulting in a greater percentage [64]. In view of elementary analysis, Luffa is mainly constituted of carbon (46.5–66.1%), oxygen (21.3–37.6%), hydrogen (6.8–8.2%), and nitrogen

Table 1 Summary of heavy metal, dyes, and pharmaceutical adsorption onto chitosan-based adsorbents

No.	Adsorbent	Adsorbate	Adsorption capacity (mg/g)	Refs.
1	Ninhydrin-functionalized chitosan	Pb (II)	196	[41]
2	Alginate/melamine/chitosan aerogel	Pb (II)	1331.6	[42]
3	Melamine-grafted chitosan	Pb (II) Hg (II)	618.2 490.7	[43]
4	Chitosan-cellulose magnetic carbon foam	Cu (II)	115.65	[44]
5	Hydroxypropyl chitosan-based dual self-healing hydrogel	Cr (VI)	95.31	[45]
6	Chitosan–montmorillonite/polyaniline nanocomposite	Methylene blue	111	[46]
7	Chitosan-deep eutectic solvents beads	Malachite green	17.86	[47]
8	Modified Chitosan Magnetic Composite	Methyl orange	785	[48]
9	Chitosan-activated carbon composite microspheres	Methyl orange	35.4	[49]
10	Xanthated chitosan/cellulose sponges	Methylene blue Congo red	213.220 289.855	[50]
11	PAMAM grafted chitosan dendrimers	Amido Black 10B Congo red	543.20 557.26	[51]
12	Cross-linked chitosan/bentonite	Amido Black 10B	350.9	[52]
13	Zr loaded cross-linked chitosan/bentonite composite	Amido Black 10B	418.4	[53]
14	Trimethyl Chitosan/Siloxane-Hybrid Coated Fe_3O_4 Nanoparticles	Sulfamethoxazole	597.9, 24.15	[54]
15	Co-polymerized chitosan/Fe_3O_4 composite	Tetracycline	67.1	[55]
16	Genipin-crosslinked chitosan/graphene oxide-SO_3H	Tetracycline	556.28	[56]
17	Chitosan/biochar	Ciprofloxacin	82	[57]
18	$CoFe_2O_4$/activated carbon@chitosan	Ciprofloxacin	93.5	[58]
19	Al(III)-chelated cryogels of N-(2-carboxyethyl)chitosan	Ciprofloxacin	390	[59]
20	Glutaraldehyde cross-linked magnetic chitosan	Diclofenac	57.5	[60]

(3–5%) [65]. In another analysis, it demonstrated that Luffa has 4.2% protein, 1.08% lipid, 1.04% ash, 55.78% fibre, and 37.81% carbohydrate [66].

When the fruit is immature and young, it is edible/eaten as a vegetable, and as it grows and becomes ripened, a fibrous structure is formed in the plant [67]. The

Fig. 2 **a** and **b** are images of Luffa

outstanding properties of Luffa, including fibrous matrix, mechanical strength, abundancy, non-toxicity, natural-based and environmentally-friendly biomaterial, fast growth and cost-effectiveness, have made it a promising material having extensive applications [13, 68]. Luffa has been employed in dishwashing, cleaning automobiles and glassware, insulations, and stuffing for pillows, hat manufacture, automobile windshield wipers, pot holders and gloves [64]. In the case of water and wastewater technology, Luffa has been continuously used for the removal of contaminants in the last two decades.

The present literature reveals that Luffa and its related composites have been used for the removal of dyes, metals, pharmaceuticals, etc. The capability of the Luffa for pollutant removal may be attributed to the surface structure of Luffa. Firstly, it is a porous biomaterial with a complex network that could provide vacant sites for the adsorption of contaminants (mainly macromolecules) via physical sorption. Secondly, it is a lignocellulosic-based material; and the cellulose surface becomes negative in contact with aqueous media, resulting in an appropriate media for the adsorption of positively charged molecules/ions [69]. In fact, the Luffa surface exhibited superior adsorption capability for anionic species. Accordingly, many scholars have attempted to exploit the mentioned advantage of Luffa for water purification.

Cadmium removal from an aqueous solution was studied by raw Luffa [70]. The adsorbent did not go under special chemical/physical treatment. It was merely washed several times with water to remove undesired dust or impurities. The experiments were carried out to optimize the adsorption capacity. It was observed that the maximum cadmium (II) elimination was occurred at neural pH 7. At low pH, a competition between Cd^{+2} and H^+ has happened, so some vacant sites are occupied by H^+ ions. At basic conditions (pH > 8), Cd^{+2} tends to precipitate which is meaningless in terms of adsorption. The equilibrium data were modeled with Freundlich and

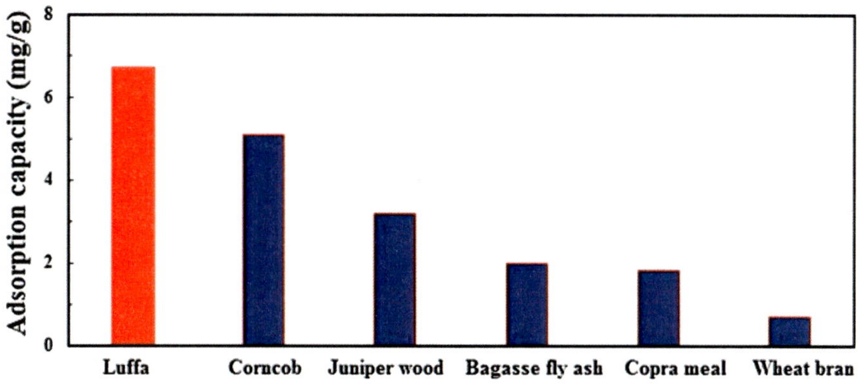

Fig. 3 Adsorption capacity of different sorbents for cadmium adsorption

Langmuir equations, and the results declared that Cd^{+2} removal follows Langmuir assumptions. The maximum adsorption capacity was reported to be 6.711 mg/g. To figure out the applicability of Luffa toward Cd^{+2}, Fig. 3 is prepared, showing the adsorption capacities of other sorbents for cadmium elimination. It is fair to suggest that Luffa had acceptable results for cadmium removal. In another study, copper removal was investigated by raw Luffa, and it was found that this biomaterial has an excellent adsorption efficiency (98%) and adsorption capacity (47.39 mg/g) [71]. Luffa was also used for the removal of a basic dye, namely, malachite green ($C_{23}H_{25}N_2$, 364.911 g/mol) [13]. It is important to note that many researchers [14–16] have already worked on this dye removal due to its significant pollution in the ecosystem. It is used to dye materials such as wool, cotton, paper and leather. In the view of Luffa utilization, the authors noted that by increasing the contact time, the effectiveness of the adsorption process enhanced until it reached the equilibrium state at 5 h. Three most-used isotherms, namely, Langmuir, Freundlich and Dubinin-Radushkevich were employed to assess the experimental results at three different temperatures. Based on the R^2 values, Langmuir (0.993) fitted the data better than Freundlich (0.867) and Dubinin-Radushkevich (0.924) isotherms. Also, the R_L values were all lower than one, revealing the appropriateness of the Langmuir model. Herein, it can be stated that malachite green molecules tend to form a monolayer on the surface of the raw Luffa. The results demonstrated that higher temperature was more favorable in terms of dye removal, and maximum monolayer sorption capacity happened at 308 K (29.4 mg/g). In short, raw Luffa could be nominated as a successful biomaterial for water purification, especially for pollutants having positive species such as heavy metals.

Although Luffa has exhibited reasonable results as a natural material for water purification, it is still not widely accepted that it could be employed for any pollutant removal. For instance, Khadir et al. expressed that raw Luffa could not be a good biomaterial for ibuprofen removal, since the results did not exhibit the required adsorption capacity [13]. The negative nature of both bio-adsorbent and adsorbate was introduced as the main reason for the adsorption process hindrance. Furthermore, many researchers are still working on Luffa to modify its capability and performance to have higher adsorption capacity and removal efficiency. To achieve this aim, composite preparation has always been expressed as a nice method of improvement, especially for natural sources.

Khadir et al. compared raw Luffa with Luffa/conducting polymer nanocomposite for the removal of ibuprofen from aqueous media [13]. Accordingly, Luffa was bought from a local shop and then washed several times with water. To conduct the polymerization step, Luffa was soaked in the solution of $FeCl_3$ (oxidant) and pyrrole (monomer). Figure 4a shows the effect of contact time on ibuprofen adsorption. At the initial step of the process, the curve owns a rapid slope, and then this continues until it reaches the equilibrium state at 90 min. The optimum pH value was found to be five (Fig. 4b). To explain, it is necessary to notice that ibuprofen is negatively charged at pH 5; however, Luffa/polypyrrole has (−NH) groups that under interaction with H^+ ions convert into $-NH_2^+$. Electrostatic interaction between positive–negative species resulted in an effective adsorption process which is illustrated also in Fig. 5 for better clarification. Pseudo-first order and Pseudo-second order models were applied to kinetics data, and it revealed that Pseudo-first order could not satisfactorily describe the process (due to its low R^2 value 0.9016 compared to Pseudo-second order 0.9987). This indicates a chemisorption mechanism for ibuprofen elimination. Among the isotherm equations, Langmuir was found to be the best with an extremely high R^2 value of 0.9879. On this occasion, the authors proposed that Luffa/polypyrrole surface is capable of proving both physical sorption (via its fibrous nature) and chemical sorption (via effective functional groups). The maximum adsorption capacity was 19.15 mg/g which was superior to previous studies, including Parthenium weed [72], Honey comb [73], Nonporous SiO_2 [74], and Olive waste [75].

Luffa-based activated carbon and biochar have always been a fascinating option for water purification. One of the first utilization of Luffa-activated carbon dates back to 2007 for the removal of reactive orange [76]. Chemical reagents such as H_3PO_4, $ZnCl_2$, and HNO_3 as well as carbonization/activation temperature of 500 °C were considered during the activated carbon preparation. The maximum adsorption capacity was 50 mg/g, and the authors believed that the obtained results are

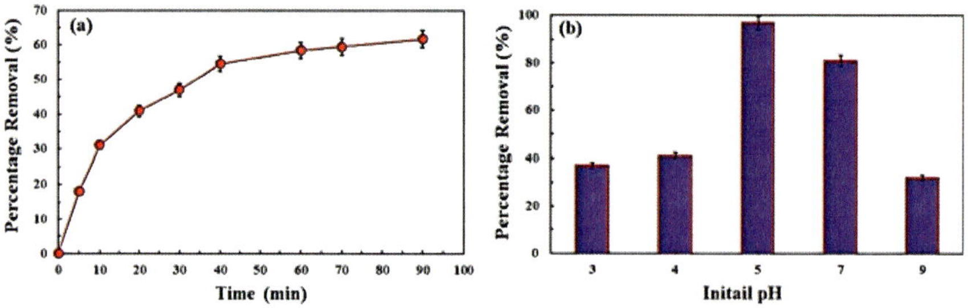

Fig. 4 Effect of contact time (**a**) and initial solution pH (**b**) on ibuprofen adsorption by Luffa/polypyrrole nanocomposite

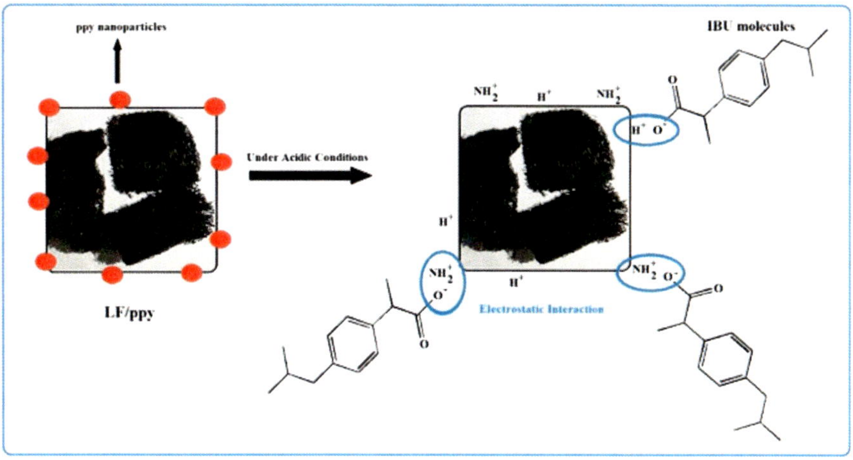

Fig. 5 Proposed mechanism for ibuprofen adsorption by Luffa/polypyrrole nanocomposite. Reprinted with permission from Elsevier and Ref. [13]

helpful for the design of a treatment plant. Interestingly, for methylene blue removal Luffa-activated carbon exhibited a significant performance by reaching the maximum adsorption capacity of 210.97 mg/g [77]. The authors tested the effect of different carbonization temperatures (600, 800, 1000 °C) and found that the effective surface area of the bioadsorbents was 472, 733, and 226 m^2/g with the total pore volumes of 0.304, 0.526, and 0.209 cm^3/g, respectively. Evidently, Luffa carbonization at 800 °C had the best surface area and pore volume. As expected, activated carbon Luffa-800 °C had higher adsorption capacity (Fig. 6). The obtained equilibrium results were analyzed by Langmuir and Freundlich isotherms. Activated carbon Luffa-800 °C was described well with the Langmuir isotherm with a R^2 value of 0.959 compared to 0.859 for Freundlich isotherm. The value of n was 1.66, indicating a favorable adsorption of methylene blue since it is in the range of 1–10. The authors applied Pseudo-first order, Pseudo-second order and intra-particle diffusion model for the analysis of the kinetics data. The R^2 value of the Pseudo-second order (0.9981) was significantly close to one and was higher than Pseudo-first order (0.8384). They deduced that sharing or exchange of electrons between adsorbent/adsorbate via chemical-reactions was the main mechanism of methylene blue adsorption. Other scholars, including Ebadollahzadeh and Zabihi [78] and Han et al. [79] also confirmed that methylene blue removal by activated carbon-based composites follows Pseudo-second order assumption. Methylene blue is one of the most studied dyes and generally is found in every textile effluent, so many studies have been conducted for its removal. Table 2 compares Luffa activated carbon with other sorbents for methylene blue adsorption. Accordingly, it is fair to suggest that Luffa provided competitive results and could be an excellent biomaterial for the dye elimination.

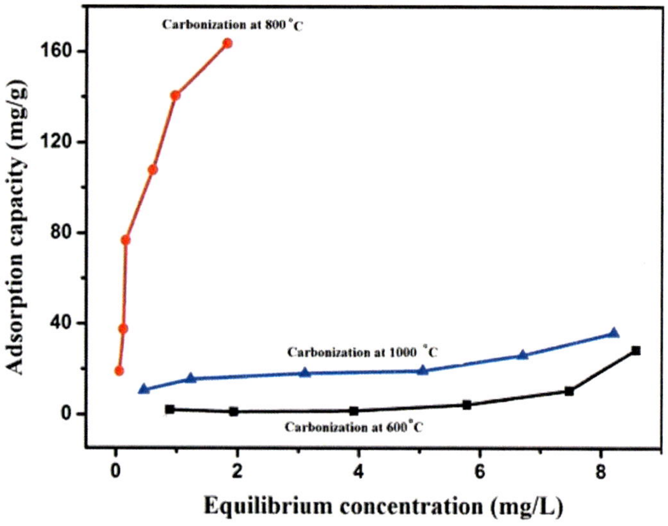

Fig. 6 Adsorption isotherms of methylene blue adsorbed by Luffa-activated carbon. Reprinted with permission from Elsevier and Ref. [77]

Table 2 The comparison of Luffa activated carbon with other materials for methylene blue adsorption in terms of adsorption capacity

No.	Adsorbent	Adsorption capacity (mg/g)	Refs.
1	Luffa activated carbon	210.97	[77]
2	Activated biochar from reed biomass	77.35	[80]
3	Pineapple waste-based activated carbon	288.34	[81]
4	Coconut waste-based activated carbon	149.30	[82]
5	Graphene oxide/calcium alginate composites	181.81	[83]
6	Eucalyptus bark-based activated carbon	204.08	[84]
7	Frozen alginate–clay	181.8	[85]
8	Magnolia denudate leaves	185.19	
9	hardwood waste "Mahogany"	149.25	[86]
10	Alginate/polyvinyl alcohol-kaolin composite	17.33	[87]
11	Platanus orientalis leaves	114.94	[88]
12	Activated carbons obtained coconut leaves	66	[89]

6.3 Sisal Fibres

Sisal (Fig. 7) belongs to the *Agave Sisalana* family, a native of Mexico, and is widely famous for its strength that made it a practical m material in the rope industry. Owning to its fantastic properties such as lightweight, robustness, inexpensiveness, and of course natural polymer, it has been successfully utilized the composite preparation [90]. Cars, furniture, construction, plastics and paper products are some of the applications of sisal. Brazil has worked so much on sisal cultivation that is currently known as the greatest producer and exporter of sisal fibre around the world. Only in Brazil, this magic fibre is capable of providing a livelihood for many people [91]. In other regions of the world such as China, Kenya, and Tanzania, this plant is cultivated, having about 1 m tall, 28 mm wide and a lifetime of about 7–10 years [92]. In order to emphasize the importance of this plant, it should be noted that in the 1960s and 1970s sisal used to be the white gold of Tanzania, contributing to the economics of this country. One of the greatest characteristics that set sisal apart from the other crops is that it could grow even in hot and arid conditions all the year. Of the main physical and mechanical properties of sisal, the following could be considered: density 1.45 g/cm^3, cellulose content 85–88%, lignin content 4–5%, tensile strength 400–700 MPa, and tensile modulus 9000–20,000 MPa [93].

To expand the applications of sisal fibre, scientists tried to use this material in water and wastewater technology. Vargas et al. milled and sieved sisal to particles from 325 to 425 μm for the removal of methylene blue and reactive black 5 dyes from aqueous solutions [94]. Even having a low surface area (5.53 m^2/g), sisal fibres represented desirable adsorption capacity which may be attributed to the functional groups in the surface of the adsorbent. Once the adsorbent dosage reached 10 g/L, almost all methylene blue and 50% of reactive black 5 were removed from the solution. In relation to the pH effect, the opposite trend was observed in which highly acidic media was favorable for reactive black 5 elimination and methylene blue had the maximum removal efficiency at approximately neutral pH. In view of thermodynamic parameters, $\Delta H°$ values were −62.60 (methylene blue) and 18.47 (reactive black 5) kJ/mol for, revealing exothermic and endothermic processes, respectively.

Fig. 7 Images of sisal plant (**a**) and sisal fibre (**b**)

The high adsorption capacity, 553.4 for methylene blue and 310.2 for reactive black 5, was one of the greatest advantages of sisal fibres that make it an excellent biomaterial for water purification. The authors clearly noted that sisal fibres could effectively remove anionic and cationic dyes from the solution. In another work regarding sisal and dye interaction, the authors prepared sisal activated carbon via chemical activation using hydrochloric acid, nitric acid, phosphoric acid, sodium hydroxide and potassium hydroxide for the removal of reactive red 120 (Fig. 8) [95]. The reagent phosphoric acid yielded better carbon than the other chemicals. The surface area of the bioadsorbents was 885 m^2/g that it was very significant and beneficial for the adsorption processes. The adsorption process reached an equilibrium state at 240 min which was almost a long time compared to the investigations conducted by Çelekli et al. [96] or Jawad et al. [97] for reactive red 120 removal. Note that even 8 h has been reported in the literature as the required time for the arrival of the equilibrium in the case of reactive red 120 elimination [98]. The maximum adsorption capacity for sisal activated carbon toward reactive red 120 was 110 mg/g, confirming the fact that it could be used well for dye adsorption.

With a view to decolorize reactive orange 5, Khadir et al. attempted to modify sisal fibres with polypyrrole/polyaniline to decolorize reactive orange 5 solution [99]. Regarding the fact that conducting polymers suffer issues such as low surface area, agglomeration, and low adsorption capacity, it is proposed to be utilized in composite materials. The following is the chief polymerization reactions:

Pyrrole polymerization:

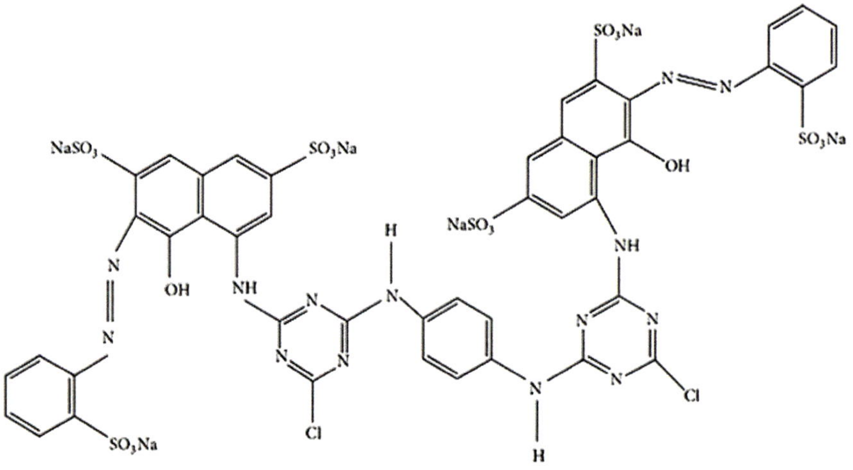

Fig. 8 The chemical structure of the reactive red 120

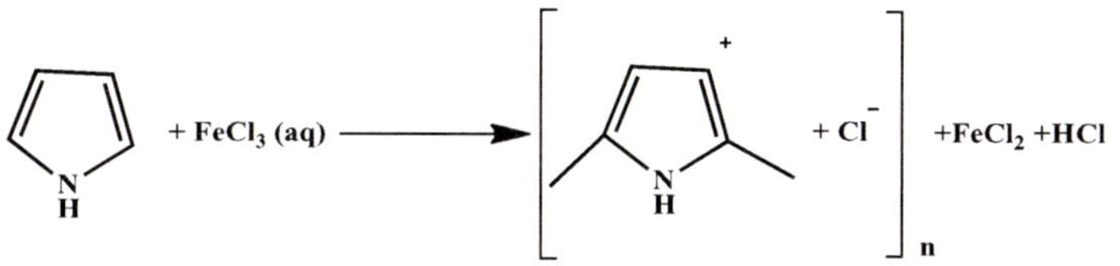

Aniline polymerization:

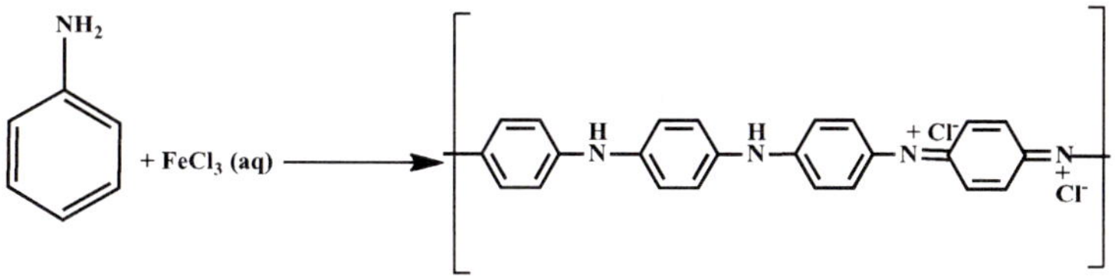

The authors firstly washed sisal fibres several times with deionized water and then conducted the polymerization procedure by use of 2 mL aniline, 2 mL pyrrole, 0.3 M HCl, and 6 g $FeCl_3$. SEM images (Fig. 9a, b) were indicative of the formation of spherical particles on the surface of raw sisal via in-situ chemical oxidative polymerization technique. It demonstrates that after modification with conducting polymers, modified sisal was more porous than the raw one. The sisal/polypyrrole/polyaniline was able to remove 68% of reactive orange 5 within 60 min of the contact time. In optimal conditions, 99% of the dye was removed from the solution. The promising performance of sisal/polypyrrole/polyaniline was ascribed to its surface functional groups that are able to establish chemical bonds with the reactive dye. Figure 9c illustrates the removal trend of reactive orange 5 by sisal and sisal/polypyrrole/polyaniline. Clearly, sisal-conducting nanoparticles composite unveiled a significantly acceptable performance than raw sisal for the dye removal, advocating that sisal fibres could act as a sustainable media for conducting polymers. It was also shown that equilibrium capacity decreased from 9.4 to 4.8 mg/g with an increase in the adsorbent dosage from 50 to 200 mg. Increment in temperature has a positive effect on the whole process. In addition, to explain the mechanism of sorption, various isotherms were applied and found that Langmuir and Freundlich were suitable for the equilibrium data. This could be related to the formation of both monolayer and multilayers of dye molecules on the heterogeneous surface of sisal/polypyrrole/polyaniline through physical and chemical reactions. The parameter n value was greater than one, indicating the dominant of chemisorption for dye adsorption on sisal composite. The following order was considered based on the isotherm suitability:

$$\text{Freundlich} > \text{Langmuir} > \text{Temkin} > \text{Dubinin} - \text{Radushkevich} > \text{Harkins} - \text{Jura}$$

In terms of kinetics models, Pseudo-first order, Pseudo-second order, Elovich, and Fractional power equation were intended and compared initially based on the R^2 values. For the kinetics models, R^2 values were in the range of 0.911–0.9921,

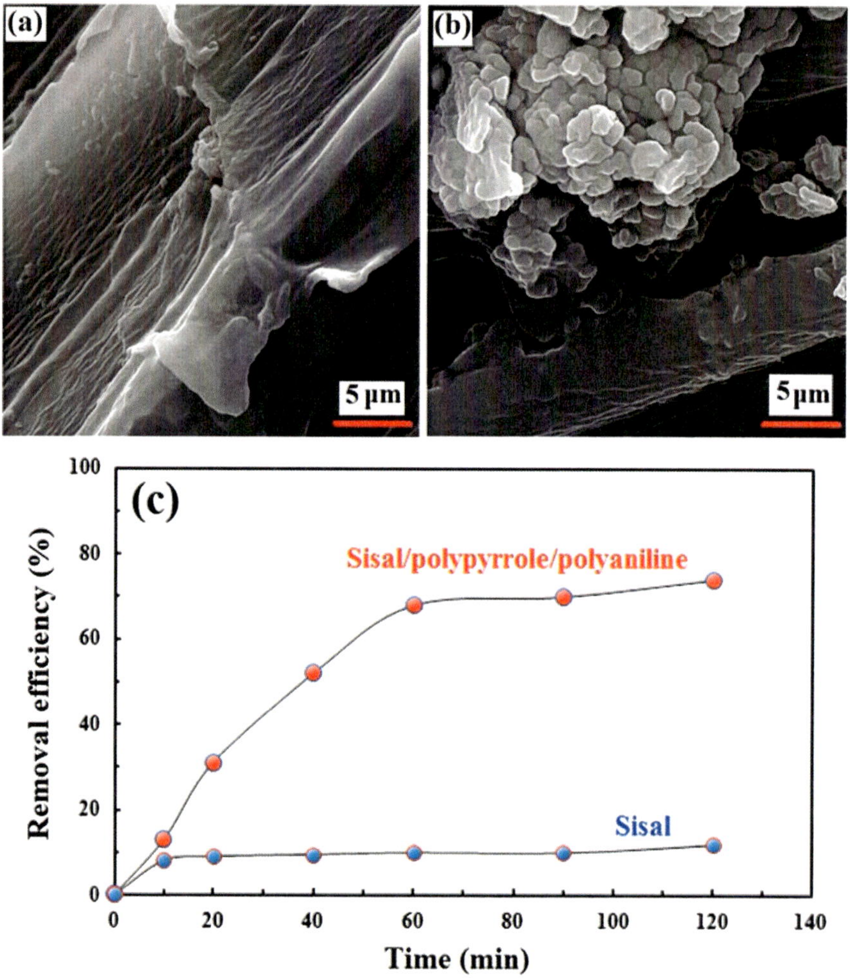

Fig. 9 SEM images of raw sisal (**a**) and sisal/polypyrrole/polyaniline (**b**). Comparison of sisal and sisal/polypyrrole/polyaniline composite for reactive orange 5 adsorption. Reprinted with permission from Elsevier and Ref. [99]

0.9934–0.9981, 0.9441–0.9898, and 0.8559–0.9525, respectively, uncovering that system followed the Pseudo-second order model. Optimization experiments stated 12.43 mg/g as the maximum adsorption capacity of sisal composite for reactive orange 5 adsorption. In another investigation, polyaniline was coated on sisal fibres for the removal of Pb (II) [100]. The authors found that the amount of Pb (II) removed sharply increased from 48 to 97.5% with an increase in adsorbent dosage from 0.1 to 1 gr, and further increase in adsorbent dosage had little effect on the amount of the removed. Within pH 4–7 the efficiency of the adsorption was almost constant and had a minor influence on Pb (II) adsorption. In EDAX analysis, the elements

O and C were detected for sisal, and after modification, other elements N and Cl were also appeared, showing a successful synthesis procedure. The Pb (II) adsorption process by sisal/polyaniline composite obeyed Langmuir assumptions with the maximum adsorption capacity of 6.53 mg/g. Polypyrrole was also used with sisal pulp for the removal of hexavalent [101]. Hexavalent removal decreased with increasing solution pH. At pH 2 the maximum removal of 99.2% and 55.6% was found for sisal pulp/polypyrrole and polypyrrole, respectively. The authors also noticed that pyrrole concentration was of great importance in the removal of adsorbate, having the optimum value of 7 g/L. The sorption kinetic data fitted well to the pseudo-second order model and isotherm data fitted well to the Langmuir isotherm model. The Langmuir maximum adsorption capacities were 326.79, 358.42 and 377.35 mg/g at 298, 308, and 318 K, respectively, with the R^2 values of 0.9988, 0.9998, and 1.000, respectively. Such high adsorption capacity may be ascribed to the presence of polypyrrole containing nitrogen groups that have a high binding affinity to metal ions. In a previous study that polypyrrole-based composite and metal ions interacted with each other in a batch system, the significant adsorption capacity was reported [102].

During the last decade, many investigations focused on the utilization of chemicals such as Fe, Zr, Al, Co, and Cr as a chemical treatment to modify the applicability of the raw materials. Fe might be the most interesting one due to its low solubility and toxicity. Fe-loaded sisal activated carbon was prepared by Mwakabona et al. for water defluoridation [103]. The authors tested the effect of post-alkalization and found that it could contribute to chemical stability and remove fluoride via electrostatic interaction. However, ligand exchange was introduced as the chief mechanism for Fe-loaded sisal activated carbon.

6.4 *Hydroxyapatite*

Hydroxyapatite, (Ca_{10} $(PO_4)_6$ $(OH)_2$), belongs to the phosphate family having the hardest structure and is found in the bones and teeth of mammals. Excellent properties of hydroxyapatite have made it a useful biomaterial in bone tissue engineering, biologicals' delivery systems, and bioactive coatings [104]. Recently, this promising biomaterial has gained much interest in the field of environmental pollution control. The presence of C and P sites in the structure of the hydroxyapatite plays the important role in removing pollutants. Hydroxyapatite can be prepared from various sources, including sea shells, sea corals, crocodile egg shells, egg shells, bamboo, fish scale and bone, pig bone, cow bone, human bone, etc. [105]. Figure 10 shows some of the main synthesis/preparation methods of hydroxyapatite.

Harja and Ciobanu proposed hydroxyapatite as a potential biomaterial for the adsorption of oxytetracycline [106]. The zwitterionic nature of the hydroxyapatite surface was proved by the value of pH_{zpc} (= 6.5) in which as pH > 6.5, = OPO_3 H^- sites are formed and at pH < 6.5 the adsorbent surface becomes negative by the formation of CaO H_2^+. Ion-exchange, surface complexation and the formation of insoluble

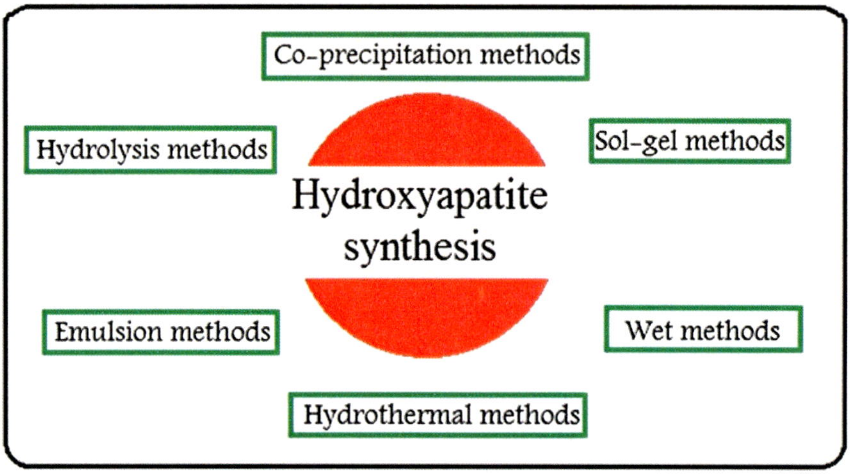

Fig. 10 Different synthesis methods for hydroxyapatite

compounds are the involved mechanisms for oxytetracycline adsorption. The whole process experienced a rapid adsorption via external surface adsorption in the first 15 min, and then reached equilibrium after an hour. The surface area 316 m^2/g and adsorption capacity 291.32 mg/g of hydroxyapatite were superior to illite/kaolinite (8.85/4.78 mg/g) [107], graphene oxide functionalized magnetic particles (45 mg/g) [108], and multiwalled carbon nanotubes (190.20 mg/g) [109]. Gao et al. prepared hierarchical hollow hydroxyapatite microspheres by the hydrothermal process for fluoride removal [110]. The kinetic data followed pseudo-second order assumptions, indicating that chemisorption is the rate-determining step in fluoride elimination by hydroxyapatite. Of studied isotherm models, Langmuir was the best model with a high R^2 value of 0.9913. The authors also surveyed the effect of competing anions and proposed that carbonate and phosphate had the most adverse effect on the performance of the adsorption, and nitrate/chloride had the least effect. This behavior was attributed to the valency of the anion. Ionexchange, electrostatic attraction, and hydrogen bonding are the primary removal pathways.

Although hydroxyapatite owns excellent property in removing pollutants, it has some drawbacks limiting its applications. Hydroxyapatite tends to agglomerate and loses dispersibility, resulting in a lower surface area [111]. In addition, the adsorption ability of hydroxyapatite can be improved by preparing suitable composites. Wei et al. suggested humic acid as organic macromolecules to improve the adsorption properties of nano-hydroxyapatite which was fabricated by a simple precipitation method [112]. In single and binary systems, the removal of Cu^{+2} and methylene blue was investigated and the optimization procedure was carried out by central composite design-based response surface methodology. At initial Cu^{+2} concentration 25 mg/L and methylene blue concentration 85 mg/L, the removal efficiencies were 97.68% and 100%, respectively (at contact time 5 h). The adsorption kinetics and equilibrium agreed well with Elovich and Langmuir models, respectively. Adsorption mechanisms were surface complexation and electrostatic attraction. In another similar study, the authors expressed the poor mechanical stability of hydroxyapatite as the main reason for modification [5]. Microwave-assisted synthesized magnetic zeolite-hydroxyapatite nanocomposite was prepared for the removal of anionic dyes. Note that such synthesis procedure could resolve the *structural* limitations of hydroxyapatite and use no surfactant or modifiers (microwave-based synthesis). Raw zeolite exhibited the BET surface area and pore volume of 35.12 m^2/g and 0.06 cm^3/g, respectively, and after modification with magnetic nanoparticles/hydroxyapatite, BET surface area and pore volume changed to 103.56 m^2/g and 0.34 cm^3/g, respectively. They found that the amount of dyes adsorbed significantly increased by reducing the solution pH, with the maximum adsorption capacity at pH 2. Thermodynamic parameters were indicative of endothermic and spontaneous reactions (negative ΔG^o and positive ΔH^o). Freundlich and pseudo-second order could satisfactorily describe the dye sorption process, demonstrating that formation of multilayers over the surface of the adsorbent via chemical-reactions is feasible. The results showed the maximum adsorption capacity of 104.05 mg/g for Congo red, 88.31 mg/g for Reactive orange 16, and 92.45 mg/g for Reactive orange 5.

Hydroxyapatite has also been used in the photocatalytic degradation of persistent pollutants. For instance, the effectiveness of low-cost red phosphorus/hollow hydroxyapatite microsphere composite (surface area 60.3 m^2/g) as a photocatalyst for degradation of rifampicin, tetracycline, and levofloxacin antibiotics [113]. The authors firstly carried out dark adsorption, meaning that adsorbents and antibiotics were interacted with each other without irradiation to ensure the establishment of adsorption–desorption equilibrium. The adsorption efficiency of 0%, 23%, and 8%, respectively. Thereafter, the reactor was illuminated by a 300 W xenon lamp, leading to complete degradation of antibiotics at 20, 10, and 50 min, respectively (pH 6, photocatalyst dosage 1 g/L). The authors proposed the following reactions for the degradation of the antibiotics:

$$\text{red phosphorus/hollow hydroxyapatite} + h\nu \rightarrow e^- + h^+$$
$$e^-(\text{hydroxyapetite}) \rightarrow e^-(\text{red phosphorus})$$
$$O_2 + e^-(\text{red phosphorus}) \rightarrow \cdot O_2^-$$
$$H_2O + h^+(\text{hydroxyapatite}) \rightarrow \cdot OH + H^+$$
$$\cdot O_2^- / \cdot OH / h^+ + \text{antibiotics} \rightarrow \text{degraded products}$$

Among various types of dyes, metal-complex dyes are considered to be highly toxic since they contain toxic complex ions as well as aromatic rings. During the removal of such dyes, the release of complex ions as the secondary pollutant must be taken into account. Heterogeneous Fenton reactions, generation of hydroxyl radicals (·OH) from reactions between recyclable iron-based catalysts and H_2O_2, has gained a great attention for its capability in degrading recalcitrant compounds. Hou et al. suggested that chitosan could be immobilized on inorganic materials such as hydroxyapatite, improving its stability, strength, and adsorptive performance [114].

Chitosan/hydroxyapatite/Fe_3O_4 magnetic composite was proposed for the removal of metal-complex dye AY220. One role of hydroxyapatite was to provide sites for adsorbing the released ions. Also, hydroxyapatite/chitosan prevents agglomeration/aggregation of bare Fe_3O_4 nanoparticles. The authors found that an increase in catalyst dosage and H_2O_2 concentration could elevate the removal efficiency, and the optimum pH was 3. The metal-complex dye AY220 removal efficiency reached 95%, and Co^{+2} ions (secondary pollutants) were effectively removed. Adsorbing Co^{+2} ions contribute to the generation of radical species and greater dye elimination. Co^{+2} released could initiate H_2O_2 via the following reactions:

$$Co^{+2}_{\text{released}} + H_2O_2 + H^+ \rightarrow Co^{+3}_{\text{released}} + \cdot OH + H_2O,$$
$$Fe^{+2} + H_2O_2 + H^+ \rightarrow Fe^{+3} + \cdot OH + H_2O.$$

However, Co^{+2} adsorbed on the surface of the catalysts experienced different reactions as follows:

$$Co^{+2}_{\text{adsorbed}} + H_2O_2 + H^+ \rightarrow Co^{+3}_{\text{adsorbed}} + \cdot OH + H_2O,$$
$$Fe^{+2} + Co^{+3}_{\text{adsorbed}} \rightarrow Fe^{+3} + Co^{+2}_{\text{adsorbed}},$$
$$Fe^{+2} + H_2O_2 + H^+ \rightarrow Fe^{+3} + \cdot OH + H_2O,$$
$$Fe^{+3} + H_2O_2 \rightarrow Fe^{+2} + HO_2 \cdot + H^+,$$
$$Fe^{+3} + HO_2 \cdot \rightarrow Fe^{+2} + O_2 + H^+.$$

Table 3 summarizes the studies conducted regarding water pollutants removal by hydroxyapatite-based adsorbents.

7 Conclusion

The findings of the chapter evidently proved that natural-based materials have gained the attention of countless scholars because of the appropriate removal efficiency of these biomaterials. They have been demonstrated to eliminate an extensive group of contaminants such as heavy metals, dyes, and pharmaceuticals. The findings of some studies are discussed and compared. The results demonstrated that biomaterials, especially those that are modified, are good candidates for wide utilization in polluted water remediation.

Declaration of Competing Interest

The authors declare that they have no known competing interests or personal relationships that could have appeared to influence the work reported in this chapter.

Table 3 Summary of the removal of various water pollutants by hydroxyapatite-based adsorbents

No.	Adsorbent	Adsorbate	Adsorption capacity (mg/g)	Refs.
1	Hydroxyapatite (eggshells)	Cu (II) Ni (II)	10.58 9.53	[115]
2	Hydroxyapatite modified sludge-based biochar	Cu (II) Cd (II)	89.98 114.68	[116]
3	Nano-hydroxyapatite	Pb (II)	1300.93	[117]
4	Mesoporous hydroxyapatite	Pb (II)	1438.85	[118]
5	Chitosan supported graphene oxide-hydroxyapatite	Congo red Acid red 1 Reactive red 2	43.06 41.32 40.03	[119]
6	Nanocrystalline hydroxyapatite	Congo red	139	[120]
7	Magnetic zeolite-hydroxyapatite	Reactive Orange 5 Reactive orange 16 Congo red	9245 88.31 104.05	[121]
8	Organofunctionalized hydroxyapatite/tripolyphosphate/chitosan spheres	Violet reactive 5R	365.0	[122]
9	Industrial waste-transformed hydroxyapatite	Congo red	139	[120]
10	Zinc (II) modified hydroxyapatites	Tetracycline	168.46	[123]
11	Hydroxyapatite	Tetracycline	5.80	[124]
12	Hydroxyapatite-clay	Tetracycline	76.02	[124]
13	Zeolite-hydroxyapatite-activated oil palm ash	Tetracycline	732.2	[125]
14	Magnetic hydroxyapatite	Catechin	110.97	[126]

References

1. Dimpe KM, Nomngongo PN (2017) A review on the efficacy of the application of myriad carbonaceous materials for the removal of toxic trace elements in the environment. Trends Environ Anal Chem 16:24–31. https://doi.org/10.1016/j.teac.2017.10.001
2. Lellis B, Fávaro-Polonio CZ, Pamphile JA, Polonio JC (2019) Effects of textile dyes on health and the environment and bioremediation potential of living organisms. Biotechnol Res Innov 3:275–290. https://doi.org/10.1016/j.biori.2019.09.001
3. Khadir A, Mollahosseini A, Tehrani RMA, Negarestani M (2020) A review on pharmaceutical removal from aquatic media by adsorption: understanding the influential parameters and novel adsorbents BT. In: Asiri AM (eds) Sustainable green chemical processes and their allied applications. Springer International Publishing, Cham, pp 207–265. https://doi.org/10.1007/978-3-030-42284-4_8

4. Beheshti F, Tehrani RMA, Khadir A (2019) Sulfamethoxazole removal by photocatalytic degradation utilizing TiO_2 and WO_3 nanoparticles as catalysts: analysis of various operational parameters. Int J Environ Sci Technol. https://doi.org/10.1007/s13762-019-02212-x
5. Piri F, Mollahosseini A, khadir A, Hosseini MM (2019) Enhanced adsorption of dyes on microwave-assisted synthesized magnetic zeolite-hydroxyapatite nanocomposite. J Environ Chem Eng 103338. https://doi.org/10.1016/j.jece.2019.103338
6. Khadir A, Negarestani M, Motamedi M (2020) Optimization of an electrocoagulation unit for purification of ibuprofen from drinking water: effect of conditions and linear/non-linear isotherm study. Sep Sci Technol. https://doi.org/10.1080/01496395.2020.1770795
7. Negarestani M, Motamedi M, Kashtiaray A, Khadir A, Sillanpää M (2020) Simultaneous removal of acetaminophen and ibuprofen from underground water by an electrocoagulation unit: operational parameters and kinetics. Groundw Sustain Dev 11100474. https://doi.org/10.1016/j.gsd.2020.100474
8. Ghenaatgar A, Tehrani RMA, Khadir A (2019) Photocatalytic degradation and mineralization of dexamethasone using WO_3 and ZrO_2 nanoparticles: optimization of operational parameters and kinetic studies. J Water Process Eng 32:100969. https://doi.org/10.1016/j.jwpe.2019.100969
9. Pal P (2020) Chapter 19—A case study on membrane-based green technology in abatement of mercury pollution. In: Pal (ed) Membrane-based technologies for environmental pollution control. Butterworth-Heinemann, pp 727–736. https://doi.org/10.1016/B978-0-12-819455-3.00019-4
10. Sun H, Yang Q, Peng Y, Shi X, Wang S, Zhang S (2010) Advanced landfill leachate treatment using a two-stage UASB-SBR system at low temperature. J Environ Sci 22:481–485. https://doi.org/10.1016/S1001-0742(09)60133-9
11. Zhao W, Chen I-W, Huang F (2019) Toward large-scale water treatment using nanomaterials. Nano Today 27:11–27. https://doi.org/10.1016/j.nantod.2019.05.003
12. Tran VS, Ngo HH, Guo W, Zhang J, Liang S, Ton-That C, Zhang X (2015) Typical low cost biosorbents for adsorptive removal of specific organic pollutants from water. Bioresour Technol 182:353–363. https://doi.org/10.1016/j.biortech.2015.02.003
13. Khadir A, Negarestani M, Mollahosseini A (2020) Sequestration of a non-steroidal anti-inflammatory drug from aquatic media by lignocellulosic material (Luffa cylindrica) reinforced with polypyrrole: study of parameters, kinetics, and equilibrium. J Environ Chem Eng 103734. https://doi.org/10.1016/j.jece.2020.103734
14. Dhabhai R, Niu CH, Dalai AK (2018) Agricultural byproducts-based biosorbents for purification of bioalcohols: a review. Bioresour Bioprocess 5:37. https://doi.org/10.1186/s40643-018-0223-7
15. Tanasă F, Teacă C-A, Nechifor M (2020) Lignocellulosic waste materials for industrial water purification BT. In: Asiri AM (ed) Sustainable green chemical processes and their allied applications. Springer International Publishing, Cham, pp 381–407. https://doi.org/10.1007/978-3-030-42284-4_14
16. Opeolu B, Bamgbose O, Arowolo T, Adetunji M (2010) Utilization of biomaterials as adsorbents for heavy metals' removal from aqueous matrices. Sci Res Essays 5:1780–1787
17. Madikizela LM, Tavengwa NT, Chimuka L (2017) Status of pharmaceuticals in African water bodies: occurrence, removal and analytical methods. J Environ Manage 193:211–220. https://doi.org/10.1016/j.jenvman.2017.02.022
18. Sabri NA, Schmitt H, Van der Zaan B, Gerritsen HW, Zuidema T, Rijnaarts HHM, Langenhoff AAM (2018) Prevalence of antibiotics and antibiotic resistance genes in a wastewater effluent-receiving river in the Netherlands. J Environ Chem Eng. https://doi.org/10.1016/j.jece.2018.03.004
19. Khadir A, Motamedi M, Pakzad E, Sillanpää M, Mahajan S (2020) The prospective utilization of Luffa fibres as a lignocellulosic bio-material for environmental remediation of aqueous media: a review. J Environ Chem Eng 104691. https://doi.org/10.1016/j.jece.2020.104691
20. Venkatesan S (2013) Adsorption. Sep Purif Technol Biorefineries 101–148. https://doi.org/10.1002/9781118493441.ch5

21. Fundamentals of the Adsorption Theory. Partit Adsorpt Org Contam Environ Syst 39–52 (2002). https://doi.org/10.1002/0471264326.ch4.
22. Desta MB (2013) Batch sorption experiments: Langmuir and Freundlich isotherm studies for the adsorption of textile metal ions onto Teff Straw (*Eragrostis tef*) agricultural waste. J Thermodyn 2013:375830. https://doi.org/10.1155/2013/375830
23. Qiu H, Lv L, Pan B, Zhang Q, Zhang W, Zhang Q (2009) Critical review in adsorption kinetic models. J Zhejiang Univ A 10:716–724
24. Largitte L, Pasquier R (2016) A review of the kinetics adsorption models and their application to the adsorption of lead by an activated carbon. Chem Eng Res Des 109:495–504. https://doi.org/10.1016/j.cherd.2016.02.006
25. Wu F-C, Tseng R-L, Huang S-C, Juang R-S (2009) Characteristics of pseudo-second-order kinetic model for liquid-phase adsorption: a mini-review. Chem Eng J 151:1–9. https://doi.org/10.1016/j.cej.2009.02.024
26. Tran HN, You S-J, Hosseini-Bandegharaei A, Chao H-P (2017) Mistakes and inconsistencies regarding adsorption of contaminants from aqueous solutions: a critical review. Water Res 120:88–116. https://doi.org/10.1016/j.watres.2017.04.014
27. Aharoni C, Ungarish M (1977) Kinetics of activated chemisorption. Part 2.—theoretical models. J Chem Soc Faraday Trans 1 Phys Chem Condens Phases 73:456–464
28. Inyang HI, Onwawoma A, Bae S (2016) The Elovich equation as a predictor of lead and cadmium sorption rates on contaminant barrier minerals. Soil Tillage Res 155:124–132. https://doi.org/10.1016/j.still.2015.07.013
29. Islam MA, Morton DW, Johnson BB, Angove MJ (2020) Adsorption of humic and fulvic acids onto a range of adsorbents in aqueous systems, and their effect on the adsorption of other species: a review. Sep Purif Technol 247116949. https://doi.org/10.1016/j.seppur.2020.116949
30. Khadir A, Motamedi M, Negarestani M, Sillanpää M, Sasani M (2020) Preparation of a nano bio-composite based on cellulosic biomass and conducting polymeric nanoparticles for ibuprofen removal: kinetics, isotherms, and energy site distribution. Int J Biol Macromol. https://doi.org/10.1016/j.ijbiomac.2020.06.095
31. Brück WM, Slater JW, Carney BF (2010) Chitin and chitosan from marine organisms. Chitin, Chitosan, Oligosaccharides Their Deriv Biol Act Appl Taylor Fr Boca Rat 11–19
32. Domard A, Domard M (2001) Chitosan: structure-properties relationship and biomedical applications. Polym Biomater 2:187–212
33. Islam S, Bhuiyan MAR, Islam MN (2017) Chitin and chitosan: structure, properties and applications in biomedical engineering. J Polym Environ 25:854–866
34. Vakili M, Deng S, Cagnetta G, Wang W, Meng P, Liu D, Yu G (2019) Regeneration of chitosan-based adsorbents used in heavy metal adsorption: a review. Sep Purif Technol 224:373–387. https://doi.org/10.1016/j.seppur.2019.05.040
35. Jiang C, Wang X, Wang G, Hao C, Li X, Li T (2019) Adsorption performance of a polysaccharide composite hydrogel based on crosslinked glucan/chitosan for heavy metal ions. Compos Part B Eng 169:45–54. https://doi.org/10.1016/j.compositesb.2019.03.082
36. Qiu W, Vakili M, Cagnetta G, Huang J, Yu G (2020) Effect of high energy ball milling on organic pollutant adsorption properties of chitosan. Int J Biol Macromol 148:543–549. https://doi.org/10.1016/j.ijbiomac.2020.01.171
37. Zhang L, Sellaoui L, Franco D, Dotto GL, Bajahzar A, Belmabrouk H, Bonilla-Petriciolet A, Oliveira MLS, Li Z (2020) Adsorption of dyes brilliant blue, sunset yellow and tartrazine from aqueous solution on chitosan: analytical interpretation via multilayer statistical physics model. Chem Eng J 382:122952. https://doi.org/10.1016/j.cej.2019.122952
38. dos Santos JMN, Pereira CR, Foletto EL, Dotto GL (2019) Alternative synthesis for $ZnFe_2O_4$/chitosan magnetic particles to remove diclofenac from water by adsorption. Int J Biol Macromol 131:301–308. https://doi.org/10.1016/j.ijbiomac.2019.03.079
39. Kevadiya BD, Rajkumar S, Bajaj HC (2015) Application and evaluation of layered silicate–chitosan composites for site specific delivery of diclofenac. Biocybern Biomed Eng 35:120–127. https://doi.org/10.1016/j.bbe.2014.08.004

40. Dehghani MH, Karri RR, Alimohammadi M, Nazmara S, Zarei A, Saeedi Z (2020) Insights into endocrine-disrupting Bisphenol-A adsorption from pharmaceutical effluent by chitosan immobilized nanoscale zero-valent iron nanoparticles. J Mol Liq 311:113317. https://doi.org/10.1016/j.molliq.2020.113317
41. Chen Y, Tang J, Wang S, Zhang L (2021) Ninhydrin-functionalized chitosan for selective removal of Pb(II) ions: characterization and adsorption performance. Int J Biol Macromol. https://doi.org/10.1016/j.ijbiomac.2021.02.110
42. Gao C, Wang X-L, An Q-D, Xiao Z-Y, Zhai S-R (2021) Synergistic preparation of modified alginate aerogel with melamine/chitosan for efficiently selective adsorption of lead ions. Carbohydr Polym 256:117564. https://doi.org/10.1016/j.carbpol.2020.117564
43. Ge H, Du J (2020) Selective adsorption of Pb(II) and Hg(II) on melamine-grafted chitosan. Int J Biol Macromol 162:1880–1887. https://doi.org/10.1016/j.ijbiomac.2020.08.070
44. Zhang Z, Li H, Li J, Li X, Wang Z, Liu X, Zhang L (2019) A novel adsorbent of core-shell construction of chitosan-cellulose magnetic carbon foam: synthesis, characterization and application to remove copper in wastewater. Chem Phys Lett 731:136573. https://doi.org/10.1016/j.cplett.2019.07.001
45. Cao J, He G, Ning X, Wang C, Fan L, Yin Y, Cai W (2021) Hydroxypropyl chitosan-based dual self-healing hydrogel for adsorption of chromium ions. Int J Biol Macromol 174:89–100. https://doi.org/10.1016/j.ijbiomac.2021.01.089
46. Minisy IM, Salahuddin NA, Ayad MM (2021) Adsorption of methylene blue onto chitosan–montmorillonite/polyaniline nanocomposite. Appl Clay Sci 203:105993. https://doi.org/10.1016/j.clay.2021.105993
47. Sadiq AC, Rahim NY, Suah FBM (2020) Adsorption and desorption of malachite green by using chitosan-deep eutectic solvents beads. Int J Biol Macromol 164:3965–3973. https://doi.org/10.1016/j.ijbiomac.2020.09.029
48. Yang D, Qiu L, Yang Y (2016) Efficient adsorption of methyl orange using a modified chitosan magnetic composite adsorbent. J Chem Eng Data 61:3933–3940. https://doi.org/10.1021/acs.jced.6b00706
49. Chen X, He L (2017) Microwave irradiation assisted preparation of chitosan composite microsphere for dye adsorption. Int J Polym Sci 2017:2672597. https://doi.org/10.1155/2017/2672597
50. Xu X, Yu J, Liu C, Yang G, Shi L, Zhuang X (2021) Xanthated chitosan/cellulose sponges for the efficient removal of anionic and cationic dyes. React Funct Polym 160:104840. https://doi.org/10.1016/j.reactfunctpolym.2021.104840
51. Banisheykholeslami F, Hosseini M, Darzi GN (2021) Design of PAMAM grafted chitosan dendrimers biosorbent for removal of anionic dyes: adsorption isotherms, kinetics and thermodynamics studies. Int J Biol Macromol. https://doi.org/10.1016/j.ijbiomac.2021.02.118
52. Liu Q, Yang B, Zhang L, Huang R (2015) Adsorption of an anionic azo dye by cross-linked chitosan/bentonite composite. Int J Biol Macromol 72:1129–1135
53. Zhang L, Hu P, Wang J, Huang R (2016) Adsorption of Amido Black 10B from aqueous solutions onto Zr (IV) surface-immobilized cross-linked chitosan/bentonite composite. Appl Surf Sci 369:558–566
54. Soares SF, Fernandes T, Trindade T, Daniel-da-Silva AL (2019) Trimethyl chitosan/siloxane-hybrid coated Fe_3O_4 nanoparticles for the uptake of sulfamethoxazole from water. Molecules 24. https://doi.org/10.3390/molecules24101958
55. Zhang S, Dong Y, Yang Z, Yang W, Wu J, Dong C (2016) Adsorption of pharmaceuticals on chitosan-based magnetic composite particles with core-brush topology. Chem Eng J 304:325–334. https://doi.org/10.1016/j.cej.2016.06.087
56. Liu Y, Liu R, Li M, Yu F, He C (2019) Removal of pharmaceuticals by novel magnetic genipin-crosslinked chitosan/graphene oxide-SO_3H composite. Carbohydr Polym 220:141–148. https://doi.org/10.1016/j.carbpol.2019.05.060
57. Afzal MZ, Sun X-F, Liu J, Song C, Wang S-G, Javed A (2018) Enhancement of ciprofloxacin sorption on chitosan/biochar hydrogel beads. Sci Total Environ 639:560–569. https://doi.org/10.1016/j.scitotenv.2018.05.129

58. Malakootian M, Nasiri A, Mahdizadeh H (2018) Preparation of $CoFe_2O_4$/activated carbon@ chitosan as a new magnetic nanobiocomposite for adsorption of ciprofloxacin in aqueous solutions. Water Sci Technol 78:2158–2170
59. Privar Y, Shashura D, Pestov A, Modin E, Baklykov A, Marinin D, Bratskaya S (2019) Metal-chelate sorbents based on carboxyalkylchitosans: ciprofloxacin uptake by Cu (II) and Al (III)-chelated cryogels of N-(2-carboxyethyl) chitosan. Int J Biol Macromol 131:806–811
60. Zhang Y, Shen Z, Dai C, Zhou X (2014) Removal of selected pharmaceuticals from aqueous solution using magnetic chitosan: sorption behavior and mechanism. Environ Sci Pollut Res 21:12780–12789
61. Zhang L, Yue Y, Shi M, Tian M, Ji J, Liao X, Hu X, Chen F (2020) Dietary Luffa cylindrica (L.) Roem promotes branched-chain amino acid catabolism in the circulation system via gut microbiota in diet-induced obese mice. Food Chem 320:126648. https://doi.org/10.1016/j.foodchem.2020.126648
62. Chen Y, Yuan F, Su Q, Yu C, Zhang K, Luo P, Hu D, Guo Y (2020) A novel sound absorbing material comprising discarded luffa scraps and polyester fibers. J Clean Prod 245:118917. https://doi.org/10.1016/j.jclepro.2019.118917
63. Kocak D, Mistik SI, Akalin M, Merdan N (2015) 21—The use of Luffa cylindrica fibres as reinforcements in composites. In: O Faruk, M Sain (eds) Biofiber reinforcements in composite materials. Woodhead Publishing, pp 689–699. https://doi.org/10.1533/9781782421276.5.689
64. Saeed A, Iqbal M (2013) Loofa (Luffa cylindrica) sponge: review of development of the biomatrix as a tool for biotechnological applications. Biotechnol Prog 29:573–600. https://doi.org/10.1002/btpr.1702
65. Laidani Y, Hanini S, Mortha G, Heninia G (2012) Study of a fibrous annual plant, Luffa cylindrica for paper application part I: characterization of the vegetal, Iran. J Chem Chem Eng 31:119–129. http://www.ijcce.ac.ir/article_5935.html
66. Oladoja NA, Aboluwoye CO, Akinkugbe AO (2009) Evaluation of Loofah as a sorbent in the decolorization of basic dye contaminated aqueous system. Ind Eng Chem Res 48:2786–2794. https://doi.org/10.1021/ie801207a
67. Bera S, Mohanty K (2020) Areca nut (Areca catechu) husks and Luffa (Luffa cylindrica) sponge as microbial immobilization matrices for efficient phenol degradation. J Water Process Eng 33:100999. https://doi.org/10.1016/j.jwpe.2019.100999
68. Kong Q, He X, Shu L, Miao M (2017) Ofloxacin adsorption by activated carbon derived from luffa sponge: Kinetic, isotherm, and thermodynamic analyses. Process Saf Environ Prot 112:254–264. https://doi.org/10.1016/j.psep.2017.05.011
69. Oboh I, Aluyor E, Audu T (2011) Application of Luffa cylindrica in natural form as biosorbent to removal of divalent metals from aqueous solutions—kinetic and equilibrium study. Waste Water-Treat Reutil. https://doi.org/10.5772/16150
70. Shahidi A, Jalilnejad N, Jalilnejad E (2015) A study on adsorption of cadmium(II) ions from aqueous solution using Luffa cylindrica. Desalin Water Treat 53:3570–3579. https://doi.org/10.1080/19443994.2013.873878
71. Ahmad R, Haseeb S (2016) Kinetic, isotherm and thermodynamic studies for the removal of Pb2+ ion by a novel adsorbent Luffa acutangula (LAPR). Desalin Water Treat 57:17826–17835. https://doi.org/10.1080/19443994.2015.1088476
72. Mondal S, Aikat K, Halder G (2016) Biosorptive uptake of ibuprofen by chemically modified Parthenium hysterophorus derived biochar: equilibrium, kinetics, thermodynamics and modeling. Ecol Eng 92:158–172
73. Dubey SP, Dwivedi AD, Sillanpää M, Gopal K (2010) Artemisia vulgaris-derived mesoporous honeycomb-shaped activated carbon for ibuprofen adsorption. Chem Eng J 165:537–544
74. Goyne KW, Chorover J, Kubicki JD, Zimmerman AR, Brantley SL (2005) Sorption of the antibiotic ofloxacin to mesoporous and nonporous alumina and silica. J Colloid Interface Sci 283:160–170. https://doi.org/10.1016/j.jcis.2004.08.150
75. Baccar R, Sarrà M, Bouzid J, Feki M, Blánquez P (2012) Removal of pharmaceutical compounds by activated carbon prepared from agricultural by-product. Chem Eng J 211–212:310–317. https://doi.org/10.1016/j.cej.2012.09.099

76. Abdelwahab O (2008) Evaluation of the use of loofa activated carbons as potential adsorbents for aqueous solutions containing dye. Desalination 222:357–367. https://doi.org/10.1016/j.desal.2007.01.146
77. Li Z, Wang G, Zhai K, He C, Li Q, Guo P (2018) Methylene blue adsorption from aqueous solution by loofah sponge-based porous carbons. Colloids Surf A Physicochem Eng Asp 538:28–35. https://doi.org/10.1016/j.colsurfa.2017.10.046
78. Ebadollahzadeh H, Zabihi M (2020) Competitive adsorption of methylene blue and Pb (II) ions on the nano-magnetic activated carbon and alumina. Mater Chem Phys 248:122893. https://doi.org/10.1016/j.matchemphys.2020.122893
79. Han Q, Wang J, Goodman BA, Xie J, Liu Z (2020) High adsorption of methylene blue by activated carbon prepared from phosphoric acid treated eucalyptus residue. Powder Technol 366:239–248. https://doi.org/10.1016/j.powtec.2020.02.013
80. Wang Y, Zhang Y, Li S, Zhong W, Wei W (2018) Enhanced methylene blue adsorption onto activated reed-derived biochar by tannic acid. J Mol Liq 268:658–666. https://doi.org/10.1016/j.molliq.2018.07.085
81. Mahamad MN, Zaini MAA, Zakaria ZA (2015) Preparation and characterization of activated carbon from pineapple waste biomass for dye removal. Int Biodeterior Biodegrad 102:274–280. https://doi.org/10.1016/j.ibiod.2015.03.009
82. Jawad AH, Rashid RA, Ishak MAM, Wilson LD (2016) Adsorption of methylene blue onto activated carbon developed from biomass waste by H_2SO_4 activation: kinetic, equilibrium and thermodynamic studies. Desalin Water Treat 57:25194–25206
83. Li Y, Du Q, Liu T, Sun J, Wang Y, Wu S, Wang Z, Xia Y, Xia L (2013) Methylene blue adsorption on graphene oxide/calcium alginate composites. Carbohydr Polym 95:501–507. https://doi.org/10.1016/j.carbpol.2013.01.094
84. Afroze S, Sen TK, Ang M, Nishioka H (2016) Adsorption of methylene blue dye from aqueous solution by novel biomass Eucalyptus sheathiana bark: equilibrium, kinetics, thermodynamics and mechanism. Desalin Water Treat 57:5858–5878
85. Uyar G, Kaygusuz H, Erim FB (2016) Methylene blue removal by alginate–clay quasi-cryogel beads. React Funct Polym 106:1–7. https://doi.org/10.1016/j.reactfunctpolym.2016.07.001
86. El Hajam M, Idrissi Kandri N, Harrach A, El khomsi A, Zerouale A (2019) Adsorption of Methylene Blue on industrial softwood waste "Cedar" and hardwood waste "Mahogany": comparative study. Mater Today Proc 13:812–821. https://doi.org/10.1016/j.matpr.2019.04.044
87. Abd El-Latif M, Latif E, Elkady M, Ibrahim A, Ossman M (2010) Alginate/polyvinyl alcohol-kaolin composite for removal of methylene blue from aqueous solution in a batch stirred tank reactor. J Am Sci 6:280–292
88. Peydayesh M, Rahbar-Kelishami A (2015) Adsorption of methylene blue onto Platanus orientalis leaf powder: kinetic, equilibrium and thermodynamic studies. J Ind Eng Chem 21:1014–1019. https://doi.org/10.1016/j.jiec.2014.05.010
89. Rashid RA, Jawad AH, Azlan M, Ishak M, Kasim NN (2018) $FeCl_3$-activated carbon developed from coconut leaves: characterization and application for methylene blue removal. Sains Malaysiana 47:603–610
90. Ansari AA, Dhakad SK, Agarwal P (2020) Investigation of mechanical properties of sisal fibre and human hair reinforced with epoxy resin hybrid polymer composite. Mater Today Proc. https://doi.org/10.1016/j.matpr.2020.02.513
91. Cantalino A, Torres E, Silva M (2015) Sustainability of sisal cultivation in brazil using co-products and wastes. J Agric Sci 7. https://doi.org/10.5539/jas.v7n7p64
92. Ramesh M (2018) 9—hemp, jute, banana, kenaf, ramie, sisal fibers. In: Bunsell AR (ed) Handbook of properties of textile and technical fibres, 2nd ed. The textile institute book series. Woodhead Publishing, pp 301–325. https://doi.org/10.1016/B978-0-08-101272-7.00009-2
93. Joseph PV, Rabello MS, Mattoso LHC, Joseph K, Thomas S (2002) Environmental effects on the degradation behaviour of sisal fibre reinforced polypropylene composites. Compos Sci Technol 62:1357–1372. https://doi.org/10.1016/S0266-3538(02)00080-5

94. Vargas VH, Paveglio RR, Pauletto P de S, Salau NPG, Dotto LG (2020) Sisal fiber as an alternative and cost-effective adsorbent for the removal of methylene blue and reactive black 5 dyes from aqueous solutions. Chem Eng Commun 207:523–536. https://doi.org/10.1080/00986445.2019.1605362
95. Dev VRG, Venugopal JR, Kumar TS, Miranda LR, Ramakrishna S (2010) Agave sisalana, a biosorbent for the adsorption of Reactive Red 120 from aqueous solution. J Text Inst 101:414–422. https://doi.org/10.1080/00405000802445289
96. Çelekli A, Al-Nuaimi AI, Bozkurt H (2019) Adsorption kinetic and isotherms of Reactive Red 120 on Moringa oleifera seed as an eco-friendly process. J Mol Struct 1195:168–178. https://doi.org/10.1016/j.molstruc.2019.05.106
97. Jawad AH, Mubarak NSA, Abdulhameed AS (2020) Tunable Schiff's base-cross-linked chitosan composite for the removal of reactive red 120 dye: adsorption and mechanism study. Int J Biol Macromol 142:732–741. https://doi.org/10.1016/j.ijbiomac.2019.10.014
98. Prola LDT, Acayanka E, Lima EC, Umpierres CS, Vaghetti JCP, Santos WO, Laminsi S, Djifon PT (2013) Comparison of Jatropha curcas shells in natural form and treated by non-thermal plasma as biosorbents for removal of Reactive Red 120 textile dye from aqueous solution. Ind Crops Prod 46:328–340. https://doi.org/10.1016/j.indcrop.2013.02.018
99. Khadir A, Negarestani M, Ghiasinejad H (2020) Low-cost sisal fibers/polypyrrole/polyaniline biosorbent for sequestration of reactive orange 5 from aqueous solutions. J Environ Chem Eng 8:103956. https://doi.org/10.1016/j.jece.2020.103956
100. Gebretsadik T, Wangatia L, Alemayehu E (2018) Removal of Pb(II) from aqueous media using adsorption onto polyaniline coated sisal fibers. J Vinyl Addit Technol 25. https://doi.org/10.1002/vnl.21652
101. Tan Y, Wei C, Gong Y, Du L (2017) Adsorption of hexavalent chromium onto sisal pulp/polypyrrole composites. IOP Conf Ser Mater Sci Eng 170:12007. https://doi.org/10.1088/1757-899X/170/1/012007
102. Mollahosseini A, Khadir A, Saeidian J (2019) Core–shell polypyrrole/Fe_3O_4 nanocomposite as sorbent for magnetic dispersive solid-phase extraction of Al^{+3} ions from solutions: investigation of the operational parameters. J Water Process Eng 29:100795. https://doi.org/10.1016/j.jwpe.2019.100795
103. Mwakabona HT, Mlay HR, Van der Bruggen B, Njau KN (2019) Water defluoridation by Fe(III)-loaded sisal fibre: understanding the influence of the preparation pathways on biosorbents' defluoridation properties. J Hazard Mater 362:99–106. https://doi.org/10.1016/j.jhazmat.2018.08.088
104. Saxena V, Shukla I, Pandey LM (2019) Chapter 8—Hydroxyapatite: an inorganic ceramic for biomedical applications. In: Grumezescu V, Grumezescu A (eds) Materials for biomedical engineering. Elsevier, pp 205–249. https://doi.org/10.1016/B978-0-12-816909-4.00008-7
105. Agbeboh NI, Oladele IO, Daramola OO, Adediran AA, Olasukanmi OO, Tanimola MO (2020) Environmentally sustainable processes for the synthesis of hydroxyapatite. Heliyon 6:e03765. https://doi.org/10.1016/j.heliyon.2020.e03765
106. Harja M, Ciobanu G (2018) Studies on adsorption of oxytetracycline from aqueous solutions onto hydroxyapatite. Sci Total Environ 628–629:36–43. https://doi.org/10.1016/j.scitotenv.2018.02.027
107. Bansal OP (2013) Sorption of tetracycline, oxytetracycline, and chlortetracycline in illite and kaolinite suspensions. ISRN Environ Chem 2013:694681. https://doi.org/10.1155/2013/694681
108. Lin Y, Xu S, Jia L (2013) Fast and highly efficient tetracyclines removal from environmental waters by graphene oxide functionalized magnetic particles. Chem Eng J 225:679–685. https://doi.org/10.1016/j.cej.2013.03.104
109. Oleszczuk P, Pan B, Xing B (2009) Adsorption and desorption of oxytetracycline and carbamazepine by multiwalled carbon nanotubes. Environ Sci Technol 43:9167–9173. https://doi.org/10.1021/es901928q
110. Gao M, Wang W, Yang H, Ye B-C (2019) Hydrothermal synthesis of hierarchical hollow hydroxyapatite microspheres with excellent fluoride adsorption property. Microporous Mesoporous Mater 289:109620. https://doi.org/10.1016/j.micromeso.2019.109620

111. Li Y, Zhang Y, Zhang Y, Wang G, Li S, Han R, Wei W (2018) Reed biochar supported hydroxyapatite nanocomposite: characterization and reactivity for methylene blue removal from aqueous media. J Mol Liq 263:53–63. https://doi.org/10.1016/j.molliq.2018.04.132
112. Wei W, Han X, Zhang M, Zhang Y, Zhang Y, Zheng C (2020) Macromolecular humic acid modified nano-hydroxyapatite for simultaneous removal of Cu(II) and methylene blue from aqueous solution: experimental design and adsorption study. Int J Biol Macromol 150:849–860. https://doi.org/10.1016/j.ijbiomac.2020.02.137
113. Zou R, Xu T, Lei X, Wu Q, Xue S (2020) Novel and efficient red phosphorus/hollow hydroxyapatite microsphere photocatalyst for fast removal of antibiotic pollutants. J Phys Chem Solids 139:109353. https://doi.org/10.1016/j.jpcs.2020.109353
114. Hou P, Shi C, Wu L, Hou X (2016) Chitosan/hydroxyapatite/Fe_3O_4 magnetic composite for metal-complex dye AY220 removal: recyclable metal-promoted Fenton-like degradation. Microchem J 128:218–225
115. Ayodele O, Olusegun SJ, Oluwasina OO, Okoronkwo EA, Olanipekun EO, Mohallem NDS, Guimarães WG, Gomes BLF de M, Souza G de O, Duarte HA (2021) Experimental and theoretical studies of the adsorption of Cu and Ni ions from wastewater by hydroxyapatite derived from eggshells. Environ Nanotechnology Monit Manag 15:100439. https://doi.org/10.1016/j.enmm.2021.100439
116. Chen Y, Li M, Li Y, Liu Y, Chen Y, Li H, Li L, Xu F, Jiang H, Chen L (2021) Hydroxyapatite modified sludge-based biochar for the adsorption of Cu^{2+} and Cd^{2+}: adsorption behavior and mechanisms. Bioresour Technol 321:124413. https://doi.org/10.1016/j.biortech.2020.124413
117. Zhou C, Wang X, Song X, Wang Y, Fang D, Ge S, Zhang R (2020) Insights into dynamic adsorption of lead by nano-hydroxyapatite prepared with two-stage ultrasound. Chemosphere 253:126661. https://doi.org/10.1016/j.chemosphere.2020.126661
118. Jiang J, Long Y, Hu X, Hu J, Zhu M, Zhou S (2020) A facile microwave-assisted synthesis of mesoporous hydroxyapatite as an efficient adsorbent for Pb^{2+} adsorption. J Solid State Chem 289:121491. https://doi.org/10.1016/j.jssc.2020.121491
119. Sirajudheen P, Karthikeyan P, Ramkumar K, Meenakshi S (2020) Effective removal of organic pollutants by adsorption onto chitosan supported graphene oxide-hydroxyapatite composite: a novel reusable adsorbent. J Mol Liq 318:114200. https://doi.org/10.1016/j.molliq.2020.114200
120. Bensalah H, Younssi SA, Ouammou M, Gurlo A, Bekheet MF (2020) Azo dye adsorption on an industrial waste-transformed hydroxyapatite adsorbent: kinetics, isotherms, mechanism and regeneration studies. J Environ Chem Eng 103807
121. Piri F, Mollahosseini A, Khadir A, Milani Hosseini M (2019) Enhanced adsorption of dyes on microwave-assisted synthesized magnetic zeolite-hydroxyapatite nanocomposite. J Environ Chem Eng 7:103338. https://doi.org/10.1016/j.jece.2019.103338
122. Pereira MBB, Honório LMC, Lima-Júnior CG, Silva Filho EC, Gaslain F, Rigaud B, Fonseca MG, Jaber M (2020) Modulating the structure of organofunctionalized hydroxyapatite/tripolyphosphate/chitosan spheres for dye removal. J Environ Chem Eng 8:103980. https://doi.org/10.1016/j.jece.2020.103980
123. Oliveira C, de Oliveira ALM, Chantelle L, Landers R, Medina-Carrasco S, Del Mar Orta M, Silva Filho EC, Fonseca MG (2021) Zinc (II) modified hydroxyapatites for tetracycline removal: Zn (II) doping or ZnO deposition and their influence in the adsorption. Polyhedron 194:114879. https://doi.org/10.1016/j.poly.2020.114879
124. Ersan M, Guler UA, Acıkel U, Sarioglu M (2015) Synthesis of hydroxyapatite/clay and hydroxyapatite/pumice composites for tetracycline removal from aqueous solutions. Process Saf Environ Prot 96:22–32. https://doi.org/10.1016/j.psep.2015.04.001
125. Khanday WA, Hameed BH (2018) Zeolite-hydroxyapatite-activated oil palm ash composite for antibiotic tetracycline adsorption. Fuel 215:499–505. https://doi.org/10.1016/j.fuel.2017.11.068
126. Yusoff AHM, Salimi MN, Gopinath SCB, Abdullah MMA, Samsudin EM (2020) Catechin adsorption on magnetic hydroxyapatite nanoparticles: a synergistic interaction with calcium ions. Mater Chem Phys 241:122337. https://doi.org/10.1016/j.matchemphys.2019.122337

Chitosan—A Promising Biomaterial for Dye Elimination

Anu Mishra and Srikrishna Natarajan

Abstract Textile, leather and paper goods require a huge amount of synthetic dyes for their colouration. During colouration and other wet processing treatments, a considerable amount of dye remains in the aqueous bath as unused. This unused dye is the major constituent of effluent and a serious concern to the water pollution. In the past two decades, many biopolymers have been extensively studied for the removal of dyes, heavy metal ions and other contaminants from effluents. Due to the biodegradability, nontoxicity and outstanding chelation behaviour, chitosan and its derivatives have been considered as prospective biosorbents to capture the dissolved dyes from aqueous solutions. Further, the ability of this biopolymer to be shaped into flakes, powders, beads, gels, films and porous particles expands its practical applicability. This write-up is an attempt to highlight the relevancy of the chemical structure of chitosan for dye removal applications. The mechanism of working of chitosan and the process parameters affecting the efficiency of dye removal have also been discussed. The recent modifications carried out in the physical and chemical structure of chitosan to improve its dye adsorption capacity have also been thoroughly investigated. Moreover, some practical constraints limiting its usage for dye removal application have also been a part of this chapter.

Keywords Adsorption · Dye removal · Chitosan · Chemical modification · Crosslinking · Glutaraldehyde · Grafting · Nanostructure · Physical modification · Solubility

A. Mishra
Department of Textile Technology, Indian Institute of Carpet Technology, Chauri Road, Bhadohi Nagar Palika, Bhadohi 221401, Uttar Pradesh, India

S. Natarajan (✉)
Department of Textile Technology, Shri Guru Gobind Singhji Institute of Engineering & Technology, Vishnupuri, Nanded 431606, Maharashtra, India

© The Author(s), under exclusive license to Springer Nature Singapore Pte Ltd. 2021
S. S. Muthu and A. Khadir (eds.), *Novel Materials for Dye-containing Wastewater Treatment*, Sustainable Textiles: Production, Processing, Manufacturing & Chemistry,
https://doi.org/10.1007/978-981-16-2892-4_3

1 Introduction

The industrial setups involved in the wet processing of textiles, food, paper and leather generate an enormous amount of waste water [14]. This waste water carries many inorganic and organic contaminants. When the waste water containing these dyes is released into the environment without any treatment, it leads to a major pollution problem. Therefore, the release of waste water without any treatment draws serious environmental and public concern.

Even a very small trace of dye present in the water is highly visible and extremely undesirable. This is due to the fact that the dye molecules hinder the visibility of light through the aqueous medium [17]. In addition to this, the process of diffusion of oxygen in the water bodies is also badly affected, which makes the survival of aquatic creatures quite difficult. Besides this, the toxic effects of these contaminants on the human body are also alarming. Health complications such as skin irritation, development of tumours and various allergic problems have also been associated with the use of contaminated water.

Therefore, the remediation of dyes from the waste water becomes paramount. Several physical methods have been reported for the removal of dyes and colouring components from industrial waste water. These are ultrafiltration, electrodialysis, reverse osmosis and chemical methods such as electro floatation and electrochemical oxidation. In general, these methods are not found to be cost-effective. Some of these treatments involve the use of excessive chemicals, which also create further environmental concerns. Also, many of these treatments are not effective for all classes of dyes [84].

Among the various physical and chemical methods used for the remediation of unused dyes from waste water, adsorption is the most popular and widely accepted method. Adsorption transfers the dye from the aqueous effluent to a solid substrate. Thus, significantly decreases the concentration of the dye in the effluent. In adsorption, activated carbon (AC) is commonly used for dye removal due to its effectiveness. AC is prepared by utilizing various organic precursors such as coal, coconut shells, lignite, wood, biomass, synthetic fibres and polymers, etc. [26]. These organic precursors are subjected to heat treatments at relatively high temperatures of 800 to 1200 K under an inert environment using various inert gases. Therefore, in the process of carbonization and activation, AC with extended surface area and high capacity of adsorption is prepared. The high surface reactivity and micro-pore structure formed during the preparation of AC efficiently eliminates the dye molecules from waste water. However, some of the major concerns of using AC in dye removal applications include the high cost and difficulty in its regeneration after use. The adsorption efficiency of AC is also reduced after its regeneration [53]. Hence, research into the development of efficient adsorbents with low cost of dye removal treatment is at the forefront in the last few decades.

In general, a suitable adsorbent to be utilized commercially for the adsorption of unused dye should meet many desirable features. For example, it should be efficient in adsorbing a wide variety of dyes. It should have a high capacity of adsorption and

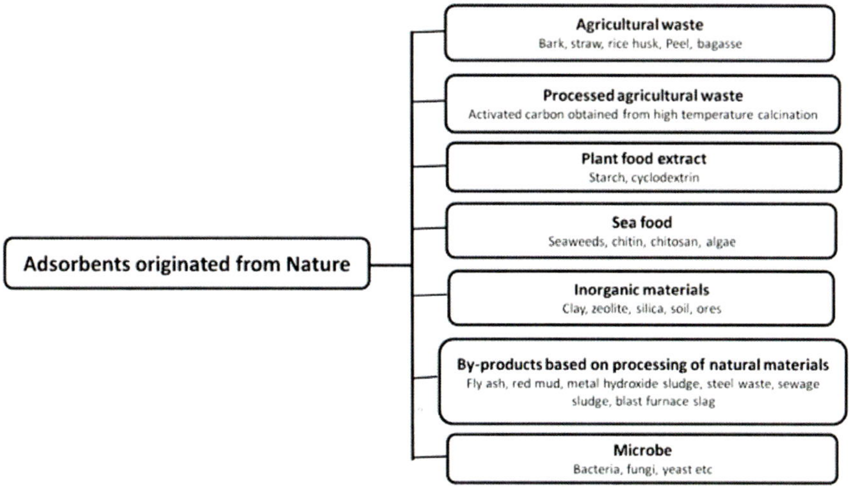

Fig. 1 Classification of adsorbents originated from nature

the rate of adsorption should also be high. It should also have a high selectivity for different concentrations of dyes. It should be structurally stable towards any adverse processing environment of waste water. In addition to this, its wide acceptability lies in the fact that it should be readily available and inexpensive [33]. Figure 1 shows the classification of adsorbents, which are originated from nature.

It has been noticed that the amount of dye adsorbed by most of the low-cost bioadsorbents is not sufficiently good. The adsorption capacity of most of the adsorbents lies in between 50 to 600 mg/g [67]. It is mainly based on the specific physical and chemical nature of the adsorbent. For example, bioadsorbent involving microorganisms may work selectively well against some dyes. However, in the rest of other cases, the dyes generally exhibit adamant behaviour towards biological degradation [39].

It is difficult to find a single adsorbent, which would have the capability to fulfill the aforementioned requirements in a comprehensive manner. Also, there is a need for screening a suitable bioadsorbent that can enhance the efficiency of the dye adsorption process. Hence, it is imperative to explore a cost-effective bioadsorbent having an improved capacity of dye adsorption. This chapter is an attempt to provide a comprehensive understanding of chitosan for its suitability in dye removal.

2 Chemical Structure and Availability

Chitosan is a biomolecule of linear polysaccharide chains. The chemical structure of chitosan is a combination of both deacetylated and acetylated units of D-glucosamine and *N*-acetyl-D-glucosamine respectively. The presence of these randomly distributed units provides heterogeneity, versatility and uniqueness in the polymer structure.

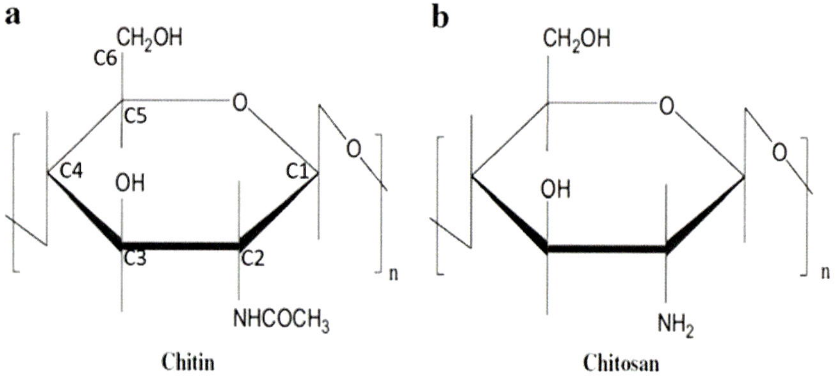

Fig. 2 Chemical structure of **a** Chitin [poly (*N*-acetyl-b-D-glucosamine)], **b** Chitosan [poly (D-glucosamine)]

Chitosan is derived from Chitin. Chitin is one of the most abundant amino-polysaccharide available in nature. Chitin is present in the exoskeleton of crustaceans, cuticles of insects and the cell walls of fungi [18]. It is a low-cost linear homopolymer, composed of β(1-4)-linked *N*-acetyl glucosamine (Fig. 2a). The chemical structure of chitin closely resembles the structure of cellulose with a slight difference. Unlike cellulose, the C-2 position of the carbon atom in chitosan is attached to acetamide in place of the hydroxyl group. The presence of the acetamide group in the chemical structure of chitosan not only improves the adsorption property of the polymer but also facilitates various modification reactions. The raw form of chitin is abundantly available as a waste of the sea food processing industry [83].

The chitin in its pure colourless form is procured from crustaceans like shrimps, crabs, etc. The acid treatment dissolves the undesired calcium carbonate from crustaceans. Subsequently, the alkaline extraction separates the proteins to obtain a colourless polymer [24]. The molecular weight of chitin typically lies between 300 and 1000 kDa, which is mainly depending upon its source of extraction.

A partial deacetylation converts chitin into chitosan (Fig. 2b). A completely deacetylized form of chitosan is not chemically stable. It gives birth to chain depolymerization and many other side reactions. Hence, a completely deacetylized form of chitosan is not regarded as a very useful polymer product. A commercial-grade chitosan is a copolymer (Fig. 3), which is characterized by its average degree of acetylation.

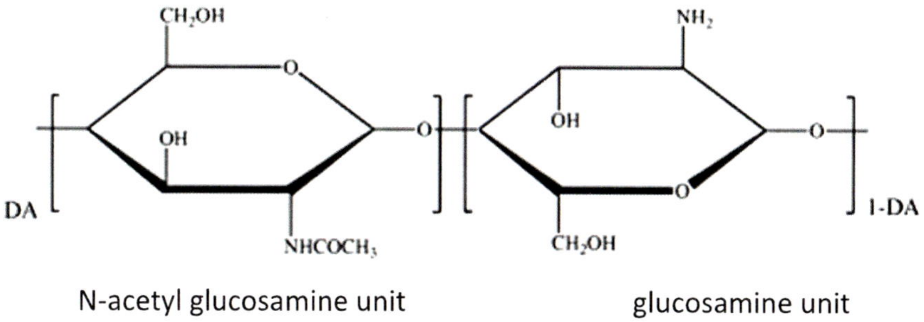

Fig. 3 Chemical structure of a commercial chitosan

3 Solubility Characteristics of Chitin and Chitosan

The crystalline structure of chitin is mainly responsible for its high solvent stability. One of the most important structural parameters on which the solubility and solution properties of both chitin and chitosan depend is the degree of acetylation (DA) or degree of deacetylation (DD). In chitin, the DA is typically around 0.90. Chitin is highly insoluble in water, aqueous solutions of dilute acids, alkalis and many organic solvents such as alcohols, hydrocarbons and ketones. However, chitin can be dissolved in some solvents such as concentrated sulphuric acid, hydrochloric acid and 85% phosphoric acid, hexafluoroacetone, hexafluoroisopropanol, chloro alcohols in conjugation with aqueous solutions of mineral acids and also in formic acid with significantly reduced molecular weights [49, 86]. For a long time, dimethyl acetamide (DMAc) containing 5% lithium chloride (LiCl) was the widely used solvent for chitin. In spite of the vast availability of chitin, recalcitrance and insolubility of chitin is a major problem that challenges the process development and application of chitin [58].

Based on the source of extraction of the biopolymer and its DD, chitin becomes soluble in aqueous acidic media. *N*-deacetylated derivative of chitin having a DD greater than 0.65 is considered as a representative chitosan molecule. Chitosan is not soluble in water at neutral pH. It is also not soluble in organic solvents like DMSO and organic alcohols. This is mainly due to its crystalline structure and presence of substantial intra and intermolecular hydrogen bonding [61]. Chitosan dissolves in organic acids (formic, acetic, propionic, succinic, toluene, acrylic, oxalic and ascorbic acids) and oxy-acids (such as gluconic, maleic, tartaric, citric and lactic acids). Chitosan can also be dissolved in acidic amino acids such as glutamic acid. This is possible due to the formation of acid salts with the existing amino groups in their structure. However, chitosan is insoluble in sulphuric, hydrochloric and phosphoric acids [80]. The most widely used solvent for chitosan is 1% acetic acid at a pH near 4. The soluble–insoluble transition of chitosan takes place at pKa value which is in between pH 6 and 6.5. Chitosan is soluble in many dilute acid solutions below pH 6. At low pH below 6, the free amino groups at the C-2 position at the polymer backbone gets protonated and becomes positively charged which makes the chitosan as a water-soluble cationic polyelectrolyte. However, at pH 6 and above, the amino groups of chitosan get deprotonated and loses its charge which makes the polymer insoluble [86]. Several factors influence the solubility of chitosan. They are mainly temperature, concentration of alkali, time of deacetylation, molar ratio of chitin to alkali solution, particle size, pretreatments applied to isolation of chitosan, distribution of the acetyl groups on the main chain and molecular weight [41, 62].

4 Utility of Chitosan in Dye Removal Application

Due to the abundant availability, the use of low-cost adsorbent from sea-based biopolymers has attracted great attention. Among them, the polysaccharides like chitin and chitosan have received considerable interest in dye adsorption. Chitosan is commercially extracted from a natural source such as crustacean (crab, crayfish, krill). Crustacean exoskeleton waste is obtained as a byproduct of the sea food processing industry. These wastes when dumped into the sea, increase the pollution in the coastal areas [15]. Hence, the utilization of these wastes to extract chitosan not only helps the economy but also saves coastal environmental pollution.

The production of chitosan does not require any exorbitant temperature; therefore, its manufacturing process is energy efficient. Due to its versatility to acquire different shapes, chitosan can be used in the form of film, fibre, sponge, gel, bead, nanocomposite and nanoparticle. Possibly, this is one of the reasons for getting a better adsorption efficiency of chitosan with different shapes vis-a-vis the conventional adsorbents.

Due to the presence of high content of both amine and hydroxyl functional groups in the chemical structure of chitosan, it has a very high affinity towards various classes of dyes. The adsorption property of chitosan has been studied for direct, acid, reactive, azo and disperse classes of dyes. However, chitosan has a low affinity towards the adsorption of cationic dyes. Only after suitable chemical modifications, the adsorption power of cationic dyes is improved.

5 Mechanism of Dye Adsorption

During adsorption, different kinds of interaction occur between dye molecules and chitosan. These include chemical bonding, ion-exchange, hydrogen bonding, hydrophobic attraction, vander Waals interaction, physical adsorption, aggregation and dye–dye interaction. For a given dye–chitosan adsorption system, these interactions may occur simultaneously or in different combinations. It is based on variation in chitosan preparation, chemical structure of dye, pH of dye effluent, salt concentrations and the type of ligands present in the dye-adsorbent system.

The removal of dye by the adsorbent surface of chitosan is carried out in the following steps:

(a) In the first step, migration of dye occurs from the bulk of the effluent dye solution to the adsorbent's surface. It is known as bulk diffusion.
(b) In the second step, diffusion of dye occurs through the boundary layer to the adsorbent's surface. It is known as film diffusion.
(c) In the third step, the dye is transported from the surface to the pores of the adsorbent. It is known as pore diffusion or intra-particle diffusion.
(d) In the fourth step, a chemical reaction occurs between the dye and the active site of the adsorbent surface. This is carried out via ion-exchange, complexation and/or chelation.

A large number of scientific work has been dedicated towards the removal of dyes using chitosan-based materials. However, most of them focus on the evaluation of the adsorption performance of the adsorbent. Only a few are aimed at exploring the mechanism of adsorption of dyes. Based on the available literature, the mechanism of adsorption of dye on chitosan is explained mainly by chemisorption, surface adsorption, aggregation and molecule exchange [36, 71].

Anionic dyes are widely studied dyes for adsorption by chitosan. It has been found that chemisorption is the most prevailing adsorption mechanism by which anionic dye molecules are trapped. In chemisorption, the adsorbate (dye molecule) is held by the adsorbent by forming chemical bonds. In this adsorption, electrons are transferred or shared between adsorbate and adsorbent. Chemisorption is a highly pH-dependent phenomenon, where adsorption takes place by ion-exchange or by electrostatic attractions. The adsorption of anionic dyes by chitosan is explained based on the chemisorption mechanism.

Under acidic conditions, the amino groups present in the chitosan chain are protonated first. At the same time, in an aqueous solution, the sulfonate group of the anionic dye is converted to anionic dye ion. In the subsequent step, the negatively charged anionic dye molecules are adsorbed on positively charged chitosan through ionic attraction [4, 38]. The rate of adsorption of anionic dyes on chitosan goes on increasing as the pH of the aqueous medium decreases.

In contrast, the usage of chitosan for the removal of cationic dyes is less preferred. As such, chitosan possesses a very low affinity towards cationic dyes due to electrostatic repulsion. Under acidic conditions, both chitosan as well as cationic dye molecule are protonated and repel each other. Hence, the adsorption is prevented. However, by some modifications in chemical structure, it is possible to utilize chitosan for the removal of cationic dyes too. Figure 4 shows the adsorption behaviour of anionic and cationic dyes with respect to chitosan under acidic conditions.

Surface adsorption is another prominent mechanism, which works to bind the dye molecules with chitosan. In this case, a surface reaction takes place. The dye molecules are attracted to the charged surface of chitosan without the exchange of ions or electrons.

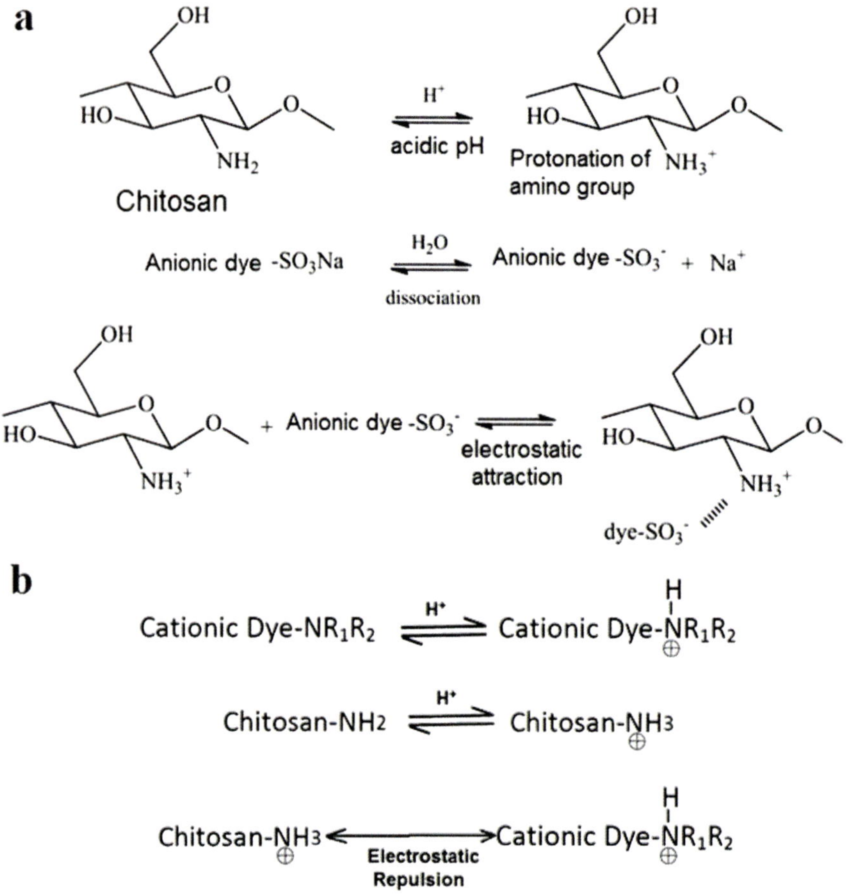

Fig. 4 Adsorption behaviour of **a** anionic and **b** cationic dyes with respect to chitosan

Sakkayawong et al. [63] studied the adsorption of a reactive dye (RR 141) by chitosan under both acidic and alkaline conditions. It has been found that the adsorption of RR 141 dye in acidic conditions followed chemisorption. However, under alkaline pH, both physical interaction (rapid surface adsorption) and chemical adsorption take place.

In another study, Uzun and Guzel [75] observed the adsorption of Orange II dye (an anionic monoazo dye of acid class) on chitosan to be of both physical and chemisorption type. It has also been concluded that physical surface interaction is a prominent mechanism in the multilayer adsorption of dyes.

Aggregation is another strong interaction that exists in the chitosan–dye adsorption system. The aggregation phenomenon is highly pH-sensitive. The mechanism of aggregation is affected by the size of the dye particles. In fact, the ability of dye particles to diffuse into the internal porous network of the adsorbent structure depends on the size of the dye particles. It has been reported that the chitosan in the form of nanoparticles can get aggregated very easily after interacting with the dye molecules [35]. Also, the concentration of dye solution influences the aggregation mechanism. A higher concentration of dye solution facilitates aggregation [20]. The aggregation can also be facilitated by adding an electrolyte to the dye solution.

It has been reported that the adsorption of dyes on grafted chitosan is a phenomenon of molecular exchange. It is not always necessary that the dye molecules shall be adsorbed by the adsorbent due to the forces of attraction. The adsorption may also take place due to strong dye–dye hydrophobic interactions, which compel the dye molecules to leave the solution and get adopted by the chitosan surface [25, 72].

Based on the studies carried out to understand the mechanism of adsorption of dye, many contrasting conclusions have been reported. It has been realized that existing simple models are not sufficient to interpret the interactions of chitosan with different classes of dyes. Therefore, more efforts are needed to explain the adsorption mechanism of chitosan–dye interactions.

6 Modifications in Chitosan

It is an established fact that amino and hydroxyl groups have a vital role in dye adsorption using chitosan. Depending upon the availability of such useful functional groups in the structure, the dye molecules in the effluent get selectively adsorbed by these adsorption sites. This is how the adsorption capacity and selectivity of chitosan alters [85, 88]. Nevertheless, modification of chitosan is desirable to improve its physical strength, solubility in acidic medium, adsorption capacity and selectivity. Various physical and chemical modifications have been made to alter the physiochemical property of chitosan to approach the requisites of an ideal adsorbent.

6.1 Modifications in Physical Form

6.1.1 Effect of the Physical Structure

The physical structure of chitosan can be modified into various forms such as powder [5], film [65], beads [55] and nanofibres [23]. This is possible either by varying the process parameters or by the adoption of various manufacturing routes. The kinetics of dye adsorption depends upon the physical form of the chitosan. A smaller size and bigger surface area of the adsorbent enhances the rate of adsorption. Similarly, higher porosity in the physical structure of the adsorbent is also helpful in increasing the adsorption rate. Eventually, the time required to achieve the equilibrium is reduced [87].

The adsorption of dye is best when chitosan is used in its dissolved polymeric form. In dissolved form, the intra-molecular H-bonding between hydroxyl and amino groups in the polymeric structure reduces. Thus, these free hydroxyl and amino groups now get more chances to interact with effluent dye molecules. However, the recovery of chitosan from its dissolved form is not easy.

On the other hand, recovery of chitosan beads in their immobilized form is easy and these beads can be reused. Wu et al. [82] studied the adsorption of RR-222 dye using chitosan flakes and beads. On comparison, it has been found that chitosan beads have a higher adsorption ability than chitosan flakes. The explanation of the obtained result lies in the fact that beads have greater surface area and porous structure as compared to the flakes. In addition to this, the residual crystallinity of polymer is considerably decreased on the formation of beads. Due to the enhancement in the specific surface and expansion in the polymer network, the diffusion property of the chitosan is improved. Therefore, using chitosan beads, the dye adsorption capacity can be more than double as compared to the flakes. Besides, it has been noticed that the chitosan converted into gel form has a reduced crystallinity. Therefore, it is possible to improve the intra-particle diffusion of dyes in its gel form. Recently, chitosan-based composites are gaining popularity. These composites offer better mechanical strength and magnetic properties [21, 47].

6.1.2 Effect of Particle Size

The particle size of the adsorbent material has a great role in determining its adsorption behaviour. Adsorption is a surface phenomenon. As the specific surface area of the adsorbent material increases, adsorption increases. The specific surface area of the adsorbent usually increases as its particle size decreases. As a result, the saturation capacity per unit weight of adsorbent increases with a decrease in particle size.

It has also been mentioned that the time elapsed in establishing the equilibrium increases, while increasing the size of the adsorbent particles. In this case, the intra-particle diffusion characteristics of the adsorbent affect the accessibility of dye to its internal sites [20, 22]. Generally, the larger size of adsorbent particles contributes more to the intra-particle diffusion resistance. This controls over the adsorption kinetics of low porosity materials.

However, in the case of large-scale applications, the use of adsorbents with small particle sizes has practical difficulties. Taking the example of a fixed-bed column, the use of small particle size frequently causes blockages in the column [73].

6.1.3 Effect of Development of Porous Structure

One of the strategies used in the modification of the physical form of chitosan is the development of the porous structure. The porous structure can be developed by the conditioning of polymer in gel form. The conditioning of gel decreases the crystallinity but improves the swelling and diffusion properties of the polymer. It also facilitates the development of porous structure in the polymer network, resulting in ease in the transportation of dyes.

By freeze-drying the solution of chitosan in acetic acid and subsequent removal of ice crystal phase by lyophilization, the material can be transformed into a porous structure. The mechanical property of the porous material is dependent on the size of pores and their orientation. By controlling the freezing rate and thermal gradients during freezing, both pore sizes and their orientation can be regulated. Alternatively, sol–gel method [16] or coagulation technique has also been used for the development of porous structure in the chitosan beads. The porous structure with high internal specific surface area allows better access of dyes to the interior adsorption sites in the polymer.

The development of nanoporous structures [66, 71, 79] with the use of nanotechnology further improves the performance due to the possibility of achieving high specific area and reduction in internal diffusion resistance as compared to the micronsized materials.

6.2 *Modifications in Chemical Structures*

Due to the abundant availability of amino and hydroxyl functional groups, the chemical modification of the chitosan structure is possible. The chemical modification in the structure not only improves the adsorption capacity of the chitosan but also imparts good mechanical resistance and chemical stability to it.

6.2.1 Modification Using Grafting

Using grafting, a desirable functional group or structure can be added to the host molecule. On the introduction of desirable functional groups, the physicochemical behaviour of the host molecule can be altered. In chitosan, grafting is possible using graft copolymerization, etherification or acylation reactions. By introducing carboxylic, amine, alkyl groups or any specific structure like crown ether or cyclodextrin, the dye adsorption performance of chitosan can be improved. The specific ligand used as a chemical graft not only improves the adsorption capacity, but also the selectivity and adsorption kinetics [81].

Graft copolymerization is one of the most popular methods of introducing desirable functional groups in chitosan. It is also used to modify the water absorption behaviour as well as the cationic/anionic properties of chitosan. The graft copolymerization reaction is initiated by free radicals, gamma irradiations or by selected enzymes [3, 40, 91]. The graft copolymerization is influenced by the type of initiator and its amount, concentration of the monomer, temperature and reaction time. Based on these parameters, the grafting percentage and grafting efficiency are determined.

In copolymerization reaction, the monomers should have reactive groups to carry out radical polymerization. Therefore, monomers are generally comprised of unsaturated structures. It has been experimentally verified that grafting of sulfonate groups in chitosan structure enhances its adsorption capacity towards cationic dyes. Similarly, grafting of amino and quaternary ammonium groups (cationic type) into chitosan improves the removal of anionic pollutants. Chitosan modified by quaternary ammonium salt can also be utilized to remove Cr(VI) from the dye effluent. Table 1 shows a modification of chitosan using grafting.

6.2.2 Modifications Using Crosslinking

Due to poor strength and chemical instability in acidic solutions, the use of unmodified chitosan is limited for the adsorption of dye from effluent. To overcome this issue, various crosslinkers have been utilized in modifying the chitosan structure [64]. Crosslinking is the formation of covalent bonds among molecular chains to form a three-dimensional network. This is possible by utilizing the functional groups present in chitosan.

The compounds having two reactive functional groups per molecule (e.g. aldehydes, anhydrides and epoxide) can be used as crosslinkers [78]. In addition to this, some special mono-functional chemical agents such as epichlorohydrin (ECH) can also be utilized as a crosslinking agent [80]. Most commonly, glutaraldehyde (GLA) [8, 56], epichlorohydrin (ECH) [46, 50] and ethylene glycol diglycidyl ether (EGDE) [28] are used for crosslinking of chitosan structure. Figure 5 shows the chemical structure of chitosan after crosslinking with GLA, ECH and EGDE respectively.

After crosslinking, the segmental mobility of the polymer chains of chitosan is restricted due to the formation of a three-dimensional network. Thus, the chemical resistance and physical strength of the adsorbent improves. Depending upon the nature and extent of crosslinking reaction, the adsorption performance of chitosan can be altered. In general, the higher the extent of crosslinking, the lower is the adsorption capacity of an adsorbent [74].

During crosslinking reaction, both hydroxyl as well as amine functional groups existing in the polymeric chain of chitosan are engaged. At the same time, crosslinking also brings the possibility of introduction of new functional groups in the adsorbent system. Therefore, modification of chitosan structure using crosslinking affects its adsorption properties to a great extent. In fact, the amino groups present in the chitosan structure participate more actively in the adsorption process as compared to hydroxyl groups. However, it is unfortunate that the crosslinkers such as GLA and EGDE react more with amino groups than hydroxyl groups. Though, crosslinker such as ECH mainly consumes hydroxyl groups rather than amino groups in crosslinking reaction. Table 2 describes the modification of chitosan using crosslinking agents.

It is practically difficult for a dye to penetrate from effluent to a highly crosslinked network structure of an adsorbent [9]. Therefore, it can easily be understood that although crosslinking is necessary for better physical resistance and chemical

Table 1 Modification of chitosan using grafting

Grafting agent	Features	Targeted class of dye/dye name	References
Succinic anhydride	• Grafting of carboxyl groups in chitosan structure. • Dye removal using electrostatic forces and chelation mechanism. • Effective in acidic medium. • Adsorbent in the form of flakes	Cationic dyes	Lima et al. [43]; Kyzas et al. [42]
Enzymatic	• Grafting of carboxyl groups in chitosan structure. • Improves the surface polarity and enhances the density of adsorption sites. • The adsorption selectivity enhances towards the targeted dye. • Adsorbent in the form of beads. • Lignin peroxidase can be immobilized on chitosan to form crosslinked enzyme aggregates (CLEA)	Cationic dyes; Reactive dyes (Sandal-fix Red C4BLN, Sandal-fix Black CKF and Sandal-fix Turq Blue GWF)	Chao et al. [7]; Parveen et al. [57]
Carboxymethylated chitosan	• Grafting of carboxyl groups. • Both physical and chemical adsorption. • Cost-effective. • Adsorbent in the form of gel. • The adsorption of congo red dye by hybrid composite using electrostatic interaction	Anionic, cationic and reactive dyes; Congo red dye	Uzun and Guzel [76, 77]; Zhang et al. [89]
Fatty acid functionalized with glycidyl moieties	• Ionic bonding and hydrophobic interaction in order to introduce long aliphatic chains	Effective adsorption of both anionic and cationic dyes.	Shimizu et al. [69]
Disulfonate derivatives	• Grafting of sulfonate groups. • Electrostatic interaction with dye. • Adsorbent in the form of bead and powder. • Sodium styrene sulfonate grafted on glutaraldehyde crosslinked microspheres	Cationic dyes; Methylene blue (MB), Methyl violet, Malachite green oxalate, safranine T and Rhodamine B	Crini and Badot [11]; Shi et al. [68]

(continued)

Table 1 (continued)

Grafting agent	Features	Targeted class of dye/dye name	References
β-cyclodextrin and graphene oxide	• Electrostatic interactions, π–π interaction and pore filling mechanism. • Adsorbent in the form of hydrogel. • Adsorption of Indigo carmine dye on chitosan crosslinked with β-cyclodextrin takes place by electrostatic interaction	Cationic dyes; Indigo carmine dye	Liu et al. [48]; Kekes and Tzia [34]

stability, a balance between the degree of crosslinking and adsorption performance of chitosan should be maintained.

6.2.3 Modifications Using Surfactant Impregnation

Surfactant impregnation is another technique by which the adsorption capability of chitosan can be improved. In this technique, the surfactant molecules are bound with chitosan using ionic links and they form supra-molecular aggregates. Surfactant impregnation facilitates the chemical linkage between hydrophilic groups of chitosan and surfactant. Besides this, the dye adsorption potential of chitosan is also improved by hydrophobic interaction between chitosan and hydrophobic part of the surfactant. Due to the strong electrostatic bonding between the macromolecules of chitosan and surfactant, the separation of surfactant from chitosan surface is not easy [6].

An anionic surfactant such as sodium dodecyl sulfate (SDS) has also been used in chitosan-based hydrogel beads to enhance the adsorption of cationic dyes. The sorption of SDS on chitosan was carried out to prepare chitosan-surfactant-core-shell beads. As compared to pure chitosan beads, the adsorption capacity of core-shell beads was almost double (360 mg/g as compared to 171.35 mg/g for removal of malachite green (MG) dye). This improvement was attributed to the "adsolubilization" of MG on the surfactant bilayer [13]. Triton X-100 and CTAB have also been used to enhance the adsorption potential of chitosan as mentioned in Table 3.

6.2.4 Modifications Using the Nanostructural Form

Due to the hydrophilicity, biodegradability and nontoxicity, chitosan is a preferred material to be utilized in nanostructural form. As compared to its micro-sized form, the nanostructure imparts high adsorption efficiency of dyes due to its size, high surface area and quantum size effect. The nanocomposite structure also shows high

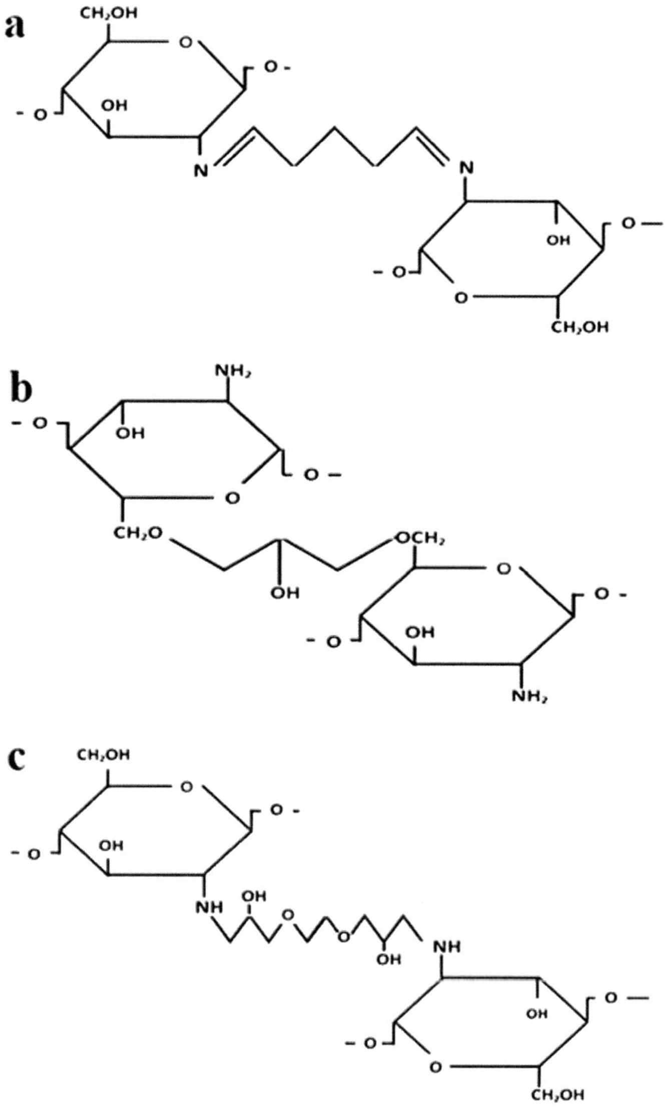

Fig. 5 Schematic representation of the crosslinked chitosan using **a** GLA, **b** ECH and **c** EGDE

stability and recyclability potential for repeated usage. Table 4 depicts some of the examples of chitosan modified into nanostructural form.

Table 2 Modification of chitosan using crosslinking agents

Crosslinking agent	Features	Targeted class of dye/ dye name	References
Glutaraldehyde (GLA)	• GLA acts as an ionic crosslinker. • It preferably crosslinks with amino groups rather than hydroxyl groups present in chitosan. • Chitosan gets crosslinked with GLA and forms bead. • Dye adsorption takes place by electrostatic interaction. • The adsorption of RB4 dye with Gluteraldehyde-crosslinked chitosan beads takes place mainly by electrostatic interaction and to a lesser extent by dipole–dipole interaction and n-pie bonds	Both cationic and anionic dyes; Reactive Blue 4 (RB4)	Kim and Cho [37]; Kamari et al. [32]; Jozwiak et al. [31]; Galan et al. [19]
Epichlorohydrin (ECH) (chloromethyloxirane)	• ECH preferably forms crosslinks with hydroxyl groups, whereas the amino adsorption sites are not much affected. • Chitosan gets crosslinked with ECH and forms bead. • Higher adsorption capacity due to unaffected amino adsorption sites, which adsorb many pollutants. • Chitosan gets crosslinked with ECH and zeolite to form zwitterionic composite beads. • Remarkable adsorptive performance for both cationic (MB dye) and anionic (RR 120 dye)	Anionic dyes; Methylene blue MB and Reactive red (RR) 120	Chiou et al. [10]; Jawad et al. [29]

(continued)

Table 2 (continued)

Crosslinking agent	Features	Targeted class of dye/ dye name	References
Ethylene glycol diglycidyl ether (EDGE)	• EGDE forms crosslink with amino groups of chitosan rather than hydroxyl groups. • The adsorption capability of crosslinked chitosan is slightly reduced. • Crosslinked chitosan forms bead. • A cost-effective process. • The adsorption of RO16 dye with chitosan/EGDE/Titanium dioxide nanoparticles works under different mechanisms such as electrostatic attraction, H-bonding, Yoshida H-bonding and n-π interaction	Acid blue 25 and Acid red 37 dyes; Reactive orange16 (RO16)	Azlan et al. [2]; [1]
Tripolyphosphate (TPP)	• Due to the polyanionic nature of TPP, electrostatic interaction takes place with the cationic part of chitosan. • Modified chitosan is in the form of nanodispersion composite and hydrogel. • Chitosan was physicochemically modified with TPP through ionic crosslinking and by fly ash particles to form a hybrid composite. • The RR120 dye adsorption mechanism is due to electrostatic attraction, Yoshida H-bonding, H-bonding and n-π bonding	Adsorption of reactive red 189 and Reactive orange 120 dyes. Also, cationic and anionic dyes such as basic blue 9, methylene blue and direct red 28	Momenzadeh et al. [52]; Li et al. [45]; Mohammed et al. [51]

(continued)

Table 2 (continued)

Crosslinking agent	Features	Targeted class of dye/ dye name	References
1,3,5-Triacryloylhexahydro-1,3,5-triazine (TAT)	Introduction of hexahydrotriazine rings of TAT into chitosan enhances the adsorption of anionic dyes	Removal of acid dyes and metal ions from solution	Shimizu et al. [70]

Table 3 Modification of chitosan using surfactant impregnation

Surfactant system	Features	Targeted class of dye/dye name	References
Impregnated with Triton X-100, a nonionic surfactant; cetyl trimethyl ammonium bromide (CTAB), a cationic surfactant; and sodium dodecyl sulfate (SDS).	• Supra-molecular aggregates of chitosan surfactant. • Complex is formed through electrostatic forces of attraction. • Direct sorption by hydrophobic interactions. • Improvement in adsorption capacity	Both cationic and anionic dyes	Chang et al. [6]
Anionic surfactant and SDS into chitosan	• Chitosan-surfactant-core-shell structure can be fabricated through sorption of SDS on chitosan. • Adsorption is improved by "adsolubilization" of cationic dye on surfactant bilayer. • It is produced in the form of hydrogel and beads	Cationic dye (Malachite Green)	Das and Pal [13]

7 Limitations of Chitosan in Dye Removal Applications

The usage of chitosan in dye removal applications is still limited at the industrial scale. Having derived from a natural source, the control over molecular weight as well as the degree of deacetylation reaction on parent polymer is difficult. Therefore, it is not easy to maintain the uniform distribution of acetyl groups along the backbone of the polymer chain. This causes variability and heterogeneity in the polymer [12]. Therefore, the reproducibility of the polymer having consistent quality remains a challenge.

Various dyes have their own complicated chemical structures. The efficiency of adsorption of dyes by chitosan shall depend upon the accessibility of their sorption sites. Besides, the uptake of dye by chitosan is highly pH-dependent [59].

Table 4 Modification of chitosan into nanostructural form

Nanocomposite system	Features	Targeted class of dye/dye name	References
Ethylenediamine-modified magnetic chitosan nanoparticles	• Separation of anionic dyes by electrostatic forces. • The force of attraction is between sulfonic anionic dyes and protonated amine of chitosan	Acid orange 7 and acid orange 10 from aqueous solutions	Zhou et al. [90]
Chitosan/magnetic maghemite (γ-Fe_2O_3) nanoparticles	• It is prepared by solution casting method	Azo dye (methyl orange) in batch adsorption experiment	Jiang et al. [30]
Hydroxyapatite by coprecipitation	• Hydroxyapatite/chitosannano composite • Ionic interaction, surface complexation, ion-exchange and H-bonding	Anionic (congo red); Brilliant green dye	Hou et al. [27]; Ragab et al. [60]
Metal oxide nanoparticle-based chitosan nanocomposite	• Chitosan layer improved the dye adsorption potential. • Due to the chelating potential of chitosan and its nanostructural form, excellent photocatalytic reduction of dye is achieved	Cationic dye (Rhodamine B); Methylene blue	Linda et al. [44]; Mostafa et al. [54]

To overcome these issues, attempts have been made to standardize various process conditions.

It is pertinent to mention that both traditional as well as commercial methods of extraction of chitin evolve a large amount of waste. The conversion of chitin to chitosan further involves a chemical deacetylation reaction, which requires a high dosage of acid.

Another issue that arises during the usage of chitosan and its derivatives is its poor physicochemical characteristics. Chitosan has a low surface area and porosity. In addition, due to strong hydrophilic characteristics and pH sensitivity, the stability of chitosan materials is generally low. Apart from this, chitosan being a biopolymer is biodegradable and does not hold durability for long-term applications.

8 Conclusion

Chitosan has been extensively studied with respect to its adsorption performance. As evident from the literature survey, chitosan has a remarkable capability of adsorbing anionic dyes such as direct, acid and reactive classes of dyes. Due to the flexible polymeric chain structure, chitosan can easily adopt various physical shapes. Therefore, as compared to others, the applicability of chitosan as an adsorbent is widely accepted. In addition to this, the enormous possibility of chemical modifications in the structure enhances its dye adsorption capacity and improves its resistance towards adverse environmental conditions.

However, it has been realized that most of the studies on adsorbent behaviour of chitosan are limited to laboratory scale batch-wise process. Therefore, pilot-plant studies should also be conducted to observe the feasibility of chitosan at a commercial scale. Besides, the adsorption property of chitosan and its derivatives should also be investigated against the industrial effluents. This would possibly be helpful in evaluating the efficacy of the adsorbent in real-life applications.

References

1. Abdulhameed AS, Jawad AH, Mohammad AT (2019) Synthesis of chitosan-ethylene glycol diglycidyl ether/TiO_2 nanoparticles for adsorption of reactive orange 16 dye using a response surface methodology approach. Bioresour Technol 293: https://doi.org/10.1016/j.biortech.2019.122071
2. Azlan K, Saime WNW, Liew LAI (2009) Chitosan and chemically modified chitosan beads for acid dyes sorption. J Environ Sci 21(3):296–302. https://doi.org/10.1016/S1001-0742(08)62267-6
3. Bahramzadeh E, Yilmaz E, Adali T (2019) Chitosan-graft-poly (N-hydroxy ethyl acrylamide) copolymers: synthesis, characterization and preliminary blood compatibility in vitro. Int J Biol Macromol 123:1257–1266. https://doi.org/10.1016/j.ijbiomac.2018.12.023
4. Binaeian E, Zadvarzi SB, Yuan D (2020) Anionic dye uptake via composite using chitosan-polyacrylamide hydrogel as matrix containing TiO_2 nanoparticles; comprehensive adsorption studies. Int J Biol Macromol 162:150–162. https://doi.org/10.1016/j.ijbiomac.2020.06.158
5. Chang YC, Chang SW, Chen DH (2006) Magnetic chitosan nanoparticles: studies on chitosan binding and adsorption of Co(II) ions. React Funct Polym 66(3):335–341. https://doi.org/10.1016/j.reactfunctpolym.2005.08.006
6. Chang Y, Jiao Y, Symons HE, Xu JF, Faul CF, Zhang X (2019) Molecular engineering of polymeric supra-amphiphiles. Chem Soc Rev 48(4):989–1003. https://doi.org/10.1039/C8CS00806J
7. Chao AC, Shyu SS, Lin YC, Mi FL (2004) Enzymatic grafting of carboxyl groups on to chitosan—to confer on chitosan the property of a cationic dye adsorbent. Bioresour Technol 91(2):157–162. https://doi.org/10.1016/S0960-8524(03)00171-8
8. Chen YW, Wang JL (2016) Removal of cesium from radioactive wastewater using magnetic chitosan beads cross-linked with glutaraldehyde. Nucl Sci Tech 27(2):1–6. https://doi.org/10.1007/s41365-016-0033-6
9. Chen M, Shen Y, Xu L, Xiang G, Ni Z (2021) Highly efficient and rapid adsorption of methylene blue dye onto vinyl hybrid silica nano-cross-linked nanocomposite hydrogel. Colloids Surf A Physicochem Eng Asp 613: https://doi.org/10.1021/acssuschemeng.9b00691
10. Chiou MS, Ho PY, Li HY (2004) Adsorption of anionic dyes in acid solutions using chemically cross-linked chitosan beads. Dyes Pigm 60(1):69–84. https://doi.org/10.1016/S0143-7208(03)00140-2
11. Crini G, Badot PM (2008) Application of chitosan, a natural aminopolysaccharide, for dye removal from aqueous solutions by adsorption processes using batch studies: a review of recent literature. Prog Polym Sci 33(4):399–447. https://doi.org/10.1016/j.progpolymsci.2007.11.001

12. Crini G, Torri G, Lichtfouse E, Kyzas GZ, Wilson LD, Morin-Crini N (2019) Cross-Linked Chitosan-based hydrogels for dye removal. In: Sustainable agriculture reviews, vol 36. Springer, Cham, pp 381–425. https://doi.org/10.1007/978-3-030-16581-9_10
13. Das D, Pal A (2016) Adsolubilization phenomenon perceived in chitosan beads leading to a fast and enhanced malachite green removal. Chem Eng J 290:371–380. https://doi.org/10.3390/app10144745
14. De Gisi S, Lofrano G, Grassi M, Notarnicola M (2016) Characteristics and adsorption capacities of low-cost sorbents for wastewater treatment: a review. SM&T 9:10–40. https://doi.org/10.1016/j.susmat.2016.06.002
15. Dhillon GS, Kaur S, Brar SK, Verma M (2013) Green synthesis approach: extraction of chitosan from fungus mycelia. Crit Rev Biotechnol 33(4):379–403. https://doi.org/10.3109/07388551.2012.717217
16. Dragan ES, Dinu MV (2020) Advances in porous chitosan-based composite hydrogels: synthesis and applications. React Funct Polym 146: https://doi.org/10.1016/j.reactfunctpolym.2019.104372
17. Dhana Rani EGS, Ganesh Kumar A, Steplinpaulselvin S, Rajaram R, Silambarasan TS, Chen Y (2020) Survival assessment of simple food webs for dye wastewater after photocatalytic degradation using SnO_2/GO nanocomposites under sunlight irradiation. Sci Total Environ 721(137805):1–9. https://doi.org/10.1016/j.scitotenv.2020.137805
18. El Knidri H, Belaabed R, Addaou A, Laajeb A, Lahsini A (2018) Extraction, chemical modification and characterization of chitin and chitosan. Int J of Biol Macromol 120:1181–1189. https://doi.org/10.1016/j.ijbiomac.2018.08.139
19. Galan J, Trilleras J, Zapata PA, Arana VA, Grande-Tovar CD (2021) Optimization of chitosan glutaraldehyde-crosslinked beads for reactive blue 4 anionic dye removal using a surface response methodology. Life 11(2):85. https://doi.org/10.3390/life11020085
20. Gibbs G, Tobin JM, Guibal E (2004) Influence of chitosan preprotonation on Reactive Black 5 sorption isotherms and kinetics. Ind Eng Chem Res 43(1):1–11. https://doi.org/10.1021/ie030352p
21. Golie WM, Upadhyayula S (2017) An investigation on biosorption of nitrate from water by chitosan based organic-inorganic hybrid biocomposites. Int J Biol Macromol 97:489–502. https://doi.org/10.1016/j.ijbiomac.2017.01.066
22. Guibal E, McCarrick P, Tobin JM (2003) Comparison of the adsorption of anionic dyes on activated carbon and chitosan derivatives from dilute solutions. Sep Sci Technol 38:3049–3073. https://doi.org/10.1081/SS-120022586
23. Habiba U, Afifi AM, Salleh A, Ang BC (2017) Chitosan/(polyvinyl alcohol)/zeolite electrospun composite nanofibrous membrane for adsorption of Cr6C, Fe3C and Ni2C. J Hazard Mater 322(Pt A):182–194. https://doi.org/10.1016/j.jhazmat.2016.06.028
24. Hahn T, Tafi E, Paul A, Salvia R, Falabella P, Zibek S (2020) Current state of chitin purification and chitosan production from insects. J Chem Technol Biotechnol 95(11):2775–2795. https://doi.org/10.1002/jctb.6533
25. Hassan MM, Carr CM (2018) A critical review on recent advancements of the removal of reactive dyes from dyehouse effluent by ion-exchange adsorbents. Chemosphere 209:201–219. https://doi.org/10.1016/j.chemosphere.2018.06.043
26. Hassan MF, Sabri MA, Fazal H, Hafeez A, Shezad N, Hussain M (2020) Recent trends in activated carbon fibers production from various precursors and applications-a comparative review. J Anal Appl Pyrolysis 145(104715):1–78. https://doi.org/10.1016/j.jaap.2019.104715
27. Hou H, Zhou R, Wu P, Wu L (2012) Removal of Congo red dye from aqueous solution with hydroxyapatite/chitosan composite. Chem Eng J 211:336–342. https://doi.org/10.1016/j.microc.2019.05.022
28. Jawad AH, Mamat NH, Hameed BH, Ismail K (2019) Biofilm of cross-linked chitosan-ethylene glycol diglycidyl ether for removal of reactive red 120 and methyl orange: adsorption and mechanism studies. J Environ Chem Eng 7(2):1–12. https://doi.org/10.1016/j.jece.2019.102965

29. Jawad AH, Abdulhameed AS, Reghioua A, Yaseen ZM (2020) Zwitterion composite chitosan-epichlorohydrin/zeolite for adsorption of methylene blue and reactive red 120 dyes. Int J Biol Macromol 163:756–765. https://doi.org/10.1016/j.ijbiomac.2020.07.014
30. Jiang R, Fu YQ, Zhu HY, Yao J, Xiao L (2012) Removal of methyl orange from aqueous solutions by magnetic maghemite/chitosan nanocomposite films: adsorption kinetics and equilibrium. J Appl Polym Sci 125(S2):E540–E549. https://doi.org/10.1002/app.37003
31. Jozwiak T, Filipkowska U, Rodziewicz J, Nowosad E (2013) Effect of cross-linking with glutaraldehyde on adsorption capacity of chitosan beads. Prog Chem Appl Chitin Deriv 18(18):35–47
32. Kamari A, Ngah WW, Chong MY, Cheah ML (2009) Sorption of acid dyes onto GLA and H_2SO_4 cross-linked chitosan beads. Desalination 249(3):1180–1189. https://doi.org/10.1016/j.desal.2009.04.010
33. Katheresan V, Kansedo J, Lau SY (2018) Efficiency of various recent wastewater dye removal methods: a review. J Environ Chem Eng 6(4):4676–4697. https://doi.org/10.1016/j.jece.2018.06.060
34. Kekes T, Tzia C (2020) Adsorption of indigo carmine on functional chitosan and β-cyclodextrin/chitosan beads: equilibrium, kinetics and mechanism studies. J Environ Manage 262: https://doi.org/10.1016/j.jenvman.2020.110372
35. Khan MSJ, Kamal T, Ali F, Asiri AM, Khan SB (2019) Chitosan-coated polyurethane sponge supported metal nanoparticles for catalytic reduction of organic pollutants. Int J Biol Macromol 132:772–783. https://doi.org/10.1016/j.ijbiomac.2019.03.205
36. Khnifira M, Boumya W, Abdennouri M, Sadiq M, Achak M, Serdaroglu G, Kaya S, Simsek S, Barka N (2021) A combined molecular dynamic simulation, DFT calculations, and experimental study of the eriochrome black T dye adsorption onto chitosan in aqueous solutions. Int J Biol Macromol 166:707–721. https://doi.org/10.1016/j.ijbiomac.2020.10.228
37. Kim TY, Cho SY (2005) Adsorption equilibria of reactive dye onto highly polyaminated porous chitosan beads. Korean J Chem Eng 22:691–696
38. Kong Q, Wang X, Lou T (2020) Preparation of millimeter-sized chitosan/carboxymethyl cellulose hollow capsule and its dye adsorption properties. Carbohydr Polym 244(116481). https://doi.org/10.1016/j.carbpol.2020.116481
39. Kumar PS, Joshiba GJ, Femina CC, Varshini P, Priyadharshini S, Karthick MA, Jothirani R (2019) A critical review on recent developments in the low-cost adsorption of dyes from wastewater. Desalin Water Treat 172:395–416. https://doi.org/10.5004/dwt.2019.24613
40. Kumar D, Gihar S, Shrivash MK, Kumar P, Kundu PP (2020) A review on the synthesis of graft copolymers of chitosan and their potential applications. Int J Biol Macromol 163:2097–2112. https://doi.org/10.1016/j.ijbiomac.2020.09.060
41. Kumari S, Annamareddy SHK, Abanti S, Rath PK (2017) Physicochemical properties and characterization of chitosan synthesized from fish scales, crab and shrimp shells. Int J Biol Macromol 104:1697–1705. https://doi.org/10.1016/j.ijbiomac.2017.04.119
42. Kyzas GZ, Siafaka PI, Pavlidou EG, Chrissafis KJ, Bikiaris DN (2015) Synthesis and adsorption application of succinyl-grafted chitosan for the simultaneous removal of zinc and cationic dye from binary hazardous mixtures. Chem Eng J 259:438–448. https://doi.org/10.1016/j.cej.2014.08.019
43. Lima IS, Ribeiro ES, Airoldi C (2006) The use of chemically modified Chitosan with succinic anhydride in the methylene blue adsorption. Quim Nova 29(3):501–506. https://doi.org/10.1590/S0100-40422006000300018
44. Linda T, Muthupoongodi S, Shajan XS, Balakumar S (2016) Fabrication and characterization of chitosan templated CdO/NiO nano composite for dye degradation. Optik 127(20):8287–8293. https://doi.org/10.1016/j.ijleo.2016.06.025
45. Li CB, Hein S, Wang K (2008) Biosorption of chitin and chitosan. Mater Sci Tech 24(9):1088. https://doi.org/10.1179/174328408x341771
46. Li CG, Cui JH, Wang F, Peng WG, He YH (2016) Adsorption removal of Congo red by epichlorohydrin-modified cross-linked chitosan adsorbent. Desalin Water Treat 57(30):14060–14066. https://doi.org/10.1080/19443994.2015.1060904

47. Liu Q, Yang BC, Zhang LJ, Huang RH (2015) Simultaneous adsorption of aniline and Cu2C from aqueous solution using activated carbon/chitosan composite. Desalin Water Treat 55(2):410–419. https://doi.org/10.1080/19443994.2014.923331
48. Liu Y, Huang S, Zhao X, Zhang Y (2018) Fabrication of three-dimensional porous β-cyclodextrin/chitosan functionalized graphene oxide hydrogel for methylene blue removal from aqueous solution. Colloids Surf A Physicochem Eng Asp 539:1–10. https://doi.org/10.1016/j.colsurfa.2017.11.066
49. Madhavan P (1992) Chitin and chitosan and their novel applications. Popular science lecture series—April Central Institute of Fisheries Technology. Cochin, India, pp 10–11
50. Marrakchi F, Hameed BH, Hummadi EH (2020) Mesoporous biohybrid epichlorohydrin crosslinked chitosan/carbon–clay adsorbent for effective cationic and anionic dyes adsorption. Int J Biol Macromol 163:1079–1086. https://doi.org/10.1016/j.ijbiomac.2020.07.032
51. Mohammed IA, Jawad AH, Abdulhameed AS, Mastuli MS (2020) Physicochemical modification of chitosan with fly ash and tripolyphosphate for removal of reactive red 120 dye: statistical optimization and mechanism study. Int J Biol Macromol 161:503–513. https://doi.org/10.1016/j.ijbiomac.2020.06.069
52. Momenzadeh H, Tehrani-Bagha AR, Khosravi A, Gharanjig K, Holmberg K (2011) Reactive dye removal from wastewater using a chitosan nanodispersion. Desalination 271(1–3):225–230. https://doi.org/10.1016/j.desal.2010.12.036
53. Momina, Shahadat M, Isamil S (2018) Regeneration performance of clay-based adsorbents for the removal of industrial dyes: a review. RSC Adv 8(43):24571–24587. https://doi.org/10.1039/C8RA04290J
54. Mostafa MH, Elsawy MA, Darwish MS, Hussein LI, Abdaleem AH (2020) Microwave-assisted preparation of Chitosan/ZnO nanocomposite and its application in dye removal. Mater Chem Phys 248: https://doi.org/10.1016/j.matchemphys.2020.122914
55. Muedas-Taipe G, Mejia IMM, Santillan FA, Velasquez CJ, Asencios YJ (2020) Removal of azo dyes in aqueous solutions using magnetized and chemically modified chitosan beads. Mater Chem Phy 256:
56. Nagireddi S, Katiyar V, Uppaluri R (2017) Pd(II) adsorption characteristics of glutaraldehyde cross-linked chitosan copolymer resin. Int J Biol Macromol 94:72–84. https://doi.org/10.1016/j.ijbiomac.2016.09.088
57. Parveen S, Asgher M, Bilal M (2021) Lignin peroxidase-based cross-linked enzyme aggregates (LiP-CLEAs) as robust biocatalytic materials for mitigation of textile dyes-contaminated aqueous solution. Environ Technol Innov 21: https://doi.org/10.1016/j.eti.2020.101226
58. Philibert T, Lee BH, Fabien N (2017) Current status and new perspectives on chitin and chitosan as functional biopolymers. Appl Biochem Biotechnol 181(4):1314–1337. https://doi.org/10.1007/s12010-016-2286-2
59. Qu B, Luo Y (2020) Chitosan-based hydrogel beads: preparations, modifications and applications in food and agriculture sectors–a review. Int J Biol Macromol 152:437–448. https://doi.org/10.1590/1678-457X.6749
60. Ragab A, Ahmed I, Bader D (2019) The removal of brilliant green dye from aqueous solution using nano hydroxyapatite/chitosan composite as a sorbent. Molecules 24(5):847. https://doi.org/10.3390/molecules24050847
61. Roy JC, Salaun F, Giraud S, Ferri A, Chen G, Guan J (2017) Solubility of chitin: solvents, solution behaviors and their related mechanisms. In: Xu Z (eds) Solubility of polysaccharides, 10. Intech open. https://doi.org/10.5772/intechopen.71385
62. Ru G, Wu S, Yan X, Liu B, Gong P, Wang L, Feng J (2019) Inverse solubility of chitin/chitosan in aqueous alkali solvents at low temperature. Carbohydr Polym 206:487–492. https://doi.org/10.1016/j.carbpol.2018.11.016
63. Sakkayawong N, Thiravetyan P, Nakbanpote W (2005) Adsorption mechanism of synthetic reactive dye wastewater by chitosan. J Colloid Interf Sci 286(1):36–42. https://doi.org/10.1016/j.jcis.2005.01.020
64. Saheed IO, Da OW, Mohd Suah FB (2021) Chitosan modifications for adsorption of pollutants: a review. J Hazard Mater 408:1–61. https://doi.org/10.1016/j.jhazmat.2020.124889

65. Salehi E, Daraei P, Shamsabadi AA (2016) A review on chitosan-based adsorptive membranes. Carbohydr Polym 152:19–432. https://doi.org/10.1016/j.carbpol.2016.07.033
66. Salehi S, Hosseinifard M (2020) Optimized removal of phosphate and nitrate from aqueous media using zirconium functionalized nanochitosan-graphene oxide composite. Cellulose 27(15):8859–8883. https://doi.org/10.1007/s10570-020-03382-5
67. Shamsollahi Z, Partovinia A (2019) Recent advances on pollutants removal by rice husk as a bio-based adsorbent: a critical review. J Environ Manage 246:314–323. https://doi.org/10.1016/j.jenvman.2019.05.145
68. Shi H, Dong C, Yang Y, Han Y, Wang F, Wang C, Men J (2020) Preparation of sulfonate chitosan microspheres and study on its adsorption properties for methylene blue. Int J Biol Macromol 163:2334–2345. https://doi.org/10.1016/j.ijbiomac.2020.09.078
69. Shimizu Y, Tanigawa S, Saito Y, Nakamura T (2005) Synthesis of chemically modified chitosans with a higher fatty acid glycidyl and their adsorption abilities for anionic and cationic dyes. J Appl Polym Sci 96(6):2423–2428. https://doi.org/10.1002/app.21707
70. Shimizu Y, Saito Y, Nakamura T (2006) Crosslinking of chitosan with a trifunctional crosslinker and the adsorption of acid dyes and metal ions onto the resulting polymer. Adsorp Sci Technol 24(1):29–39. https://doi.org/10.1260/026361706778062568
71. Sirajudheen P, Karthikeyan P, Ramkumar K, Meenakshi S (2020) Effective removal of organic pollutants by adsorption onto chitosan supported graphene oxide hydroxyapatite composite: a novel reusable adsorbent. J Mol Liq 318(114200):1–9. https://doi.org/10.1016/j.molliq.2020.114200
72. Sohni S, Hashim R, Nidaullah H, Lamaming J, Sulaiman O (2019) Chitosan/nano-lignin based composite as a new sorbent for enhanced removal of dye pollution from aqueous solutions. Int J Biol Macromol 132:1304–1317. https://doi.org/10.1016/j.ijbiomac.2019.03.151
73. Taka AL, Klink MJ, Mbianda XY, Naidoo EB (2020) Chitosan nanocomposites for water treatment by fixed-bed continuous flow column adsorption: a review. Carbohydr Polym 117398:1–34. https://doi.org/10.1016/j.carbpol.2020.117398
74. Tang T, Cao S, Xi C, Chen Z (2021) Multifunctional magnetic chitosan-graphene oxide-ionic liquid ternary nanohybrid: an efficient adsorbent of alkaloids. Carbohydr Polym 255: https://doi.org/10.1016/j.carbpol.2020.117338
75. Uzun I, Guzel F (2004) External mass transfer studies during the adsorptions of some dyestuffs and p-nitrophenol onto chitosan from aqueous solution. Turk J Chem 28(6):731–740
76. Uzun I, Guzel F (2004) Kinetics and thermodynamics of the adsorption of some dyestuffs and p-nitrophenol by chitosan and MCM-chitosan from aqueous solution. J Colloid Interface Sci 274(2):398–412. https://doi.org/10.1016/j.jcis.2004.02.022
77. Uzun I, Guzel F (2005) Rate studies on the adsorption of some dyestuffs and p-nitrophenol by chitosan and monocarboxymethylated (mcm)-chitosan from aqueous solution. J Hazard Mater 118(1–3):141–154. https://doi.org/10.1016/j.jhazmat.2004.10.006
78. Vakili M, Rafatullah M, Salamatinia B, Abdullah AZ, Ibrahim MH, Tan KB, Gholami Z, Amouzgar P (2014) Application of chitosan and its derivatives as adsorbents for dye removal from water and wastewater: a review. Carbohyd Polym 113:115–130. https://doi.org/10.1016/j.carbpol.2014.07.007
79. Vigneshwaran S, Sirajudheen P, Nabeena CP, Meenakshi S (2021) In situ fabrication of ternary TiO_2 doped grafted chitosan/hydroxyapatite nanocomposite with improved catalytic performance for the removal of organic dyes: Experimental and systemic studies. Colloids Surf A Physicochem Eng Asp 611: https://doi.org/10.1016/j.colsurfa.2020.125789
80. Wang JL, Chen C (2014) Chitosan-based biosorbents: modification and application for biosorption of heavy metals and radionuclides. Bioresour Technol 160:129–141. https://doi.org/10.1016/j.biortech.2013.12.110
81. Wang J, Zhuang S (2018) Removal of various pollutants from water and wastewater by modified chitosan adsorbents. Crit Rev Env Sci Tec 47(23):2331–2386. https://doi.org/10.1080/10643389.2017.1421845
82. Wu FC, Tseng RL, Juang RS (2001) Enhanced abilities of highly swollen chitosan beads for color removal and tyrosinase immobilization. J Hazard Mater 81(1–2):167–177. https://doi.org/10.1016/S0304-3894(00)00340-X

83. Yadav M, Goswami P, Paritosh K, Kumar M, Pareek N, Vivekanand V (2019) Seafood waste: a source for preparation of commercially employable chitin/chitosan materials. Bioresour Bioprocess 6(1):1–20. https://doi.org/10.1186/s40643-019-0243-y
84. Yeow PK, Wong SW, Hadibarata T (2021) Removal of azo and anthraquinone dye by plant biomass as adsorbent—a review. Biointerface Res Appl Chem 11(1):8218–8232. https://doi.org/10.33263/BRIAC111.82188232
85. Yu J, Zheng JD, Lu QF, Yang SX, Zhang XM, Wang X, Yang W (2016) Selective adsorption and reusability behavior for Pb2C and Cd2C on chitosan/poly(ethylene glycol)/poly(acrylic acid) adsorbent prepared by glow-discharge electrolysis plasma. Colloid Polym Sci 294(10):1585–1598. https://doi.org/10.1007/s00396-016-3920-9
86. Zargar V, Asghari M, Dashti A (2015) A review on chitin and chitosan polymers: structure, chemistry, solubility, derivatives, and applications. Chem Bio Eng Rev 2(3):204–226. https://doi.org/10.1002/cben.201400025
87. Zhang L, Zeng YX, Cheng ZJ (2016) Removal of heavy metal ions using chitosan and modified chitosan: a review. J Mol Liq 214:175–191. https://doi.org/10.1016/j.molliq.2015.12.013
88. Zhang HF, Dang QF, Liu CS, Cha DS, Yu ZZ, Zhu WJ, Fan B (2017) Uptake of Pb(II) and Cd(II) on chitosan microsphere surface successively grafted by methyl acrylate and diethylenetriamine. ACS Appl Mater Interfaces 9(12):11144–11155. https://doi.org/10.1021/acsami.7b00480
89. Zhang H, Ma J, Wang F, Chu Y, Yang L, Xia M (2020) Mechanism of carboxymethyl chitosan hybrid montmorillonite and adsorption of Pb (II) and Congo red by CMC-MMT organic-inorganic hybrid composite. Int J Biol Macromol 149:1161–1169. https://doi.org/10.1016/j.ijbiomac.2020.01.201
90. Zhou L, Jin J, Liu Z, Liang X, Shang C (2011) Adsorption of acid dyes from aqueous solutions by the ethylenediamine-modified magnetic chitosan nanoparticles. J Hazard Mater 185(2–3):1045–1052. https://doi.org/10.1016/j.jhazmat.2010.10.012
91. Zhuang S, Yin Y, Wang J (2018) Removal of cobalt ions from aqueous solution using chitosan grafted with maleic acid by gamma radiation. Nucl Eng Technol 50(1):211–215. https://doi.org/10.1016/j.jhazmat.2009.03.148

Dye Degradation by Recent Promising Composite

Djamal Zerrouki and Abdellah Henni

Abstract Dyes are employed in an assortment of manufacturing products, including textiles, paints, plastics, beverages, pharmaceuticals, and cosmetics. Dyes discharging in effluents is a big environmental concern, resulting in a high decline of water quality and destruction of the aquatic ecosystem. Various conventional routes to treat with the effluents containing dyes, such as filtration, coagulation-flocculation, and adsorption have been extensively studied, however; these processes are based only on the transportation of dyes to another phase without destroying them this is not cost-effective because an additional step to achieve a full treatment must be applied. The only way to resolve the environmental problem is to remove these dyes from effluents by degrading them into non-toxic materials. Recently the degradation of organic compounds via photocatalysis using semiconductors is considered a promising process; metal-based oxides are reported to be an efficient photocatalyst to break down the organic dyes. Here, we summarize the most up-to-date information on dye degradation using a new promising nanocomposite. Their synthesis process and photocatalytic mechanism of the hybrid structure have been addressed. The effectiveness of photocatalytic removal in comparison to traditional approaches has all been examined.

1 Introduction

The excessive utilization of dyes in many industries has dramatically increased in the last decades from drugs and cosmetic manufacture to paint, printing, plastic, food, and medical field [5, 7, 8, 16, 18, 27, 44, 47]. Organic dyes are now a fundamental part of many manufacturing processes, and are among the most significant components in the formulation process, throughout the fabrication process or the product utilization. The result of the release of dyes cause harmful soil contamination and water pollution especially. The majority of available dyes are non-biodegradable, posing a significant

D. Zerrouki (✉) · A. Henni
Laboratory of Dynamic, Interactions and Reactivity of System's, Kasdi Merbah University, 30000 Ouargla, Algeria

© The Author(s), under exclusive license to Springer Nature Singapore Pte Ltd. 2021
S. S. Muthu and A. Khadir (eds.), *Novel Materials for Dye-containing Wastewater Treatment*, Sustainable Textiles: Production, Processing, Manufacturing & Chemistry,
https://doi.org/10.1007/978-981-16-2892-4_4

threat to the atmosphere and aquatic system when released. Organic dyes are usually complex aromatic compounds with a wide range of structures. Dyes are distinct from other chemical compounds in terms of color since:

- they can absorb visible light (400–700 nm);
- contain at minimum one chromophore as colors group such as anthraquinone, methane, phthalocyanine, nitro, azo, triarylmethane;
- characterized by a conjugated system, their structure includes double and single bonds;
- expose electron resonance as a stabilizing mechanism in organic compounds.

Besides the chromophores, most coloring agents contain auxochromes, such as carboxylic acid, sulphonic acid, amino acid, and hydroxyl group, which are not qualified for coloring. But their viability can modify the color of the dye, and they are most commonly used to regulate the solubility of the dye.

To produce color in an organic compound, the chromophore must be a part of the conjugated system; the size of the conjugated system causes a major bathochromic transition.

Dyes are classified as Azo dyes, triaryl methane dyes, anthraquinone dyes, heterocyclic dyes, and phthalocyanine dyes based on their chemical structures. Azo dyes are distinguished by the inclusion of azo groups –N=N– in the molecule, which form a bridge between organic residues, up to 70% of dyes used are Azo class and represent the most important chemical class of dyes [3].

The most recent advancement in the field of water treatment has been completed in the oxidation of organic compounds, strongly present in the effluents of textile waste for example. These approaches are founded on the establishment of very reactive chemical compounds that transforms the most recalcitrant entities into degradable molecules or mineral molecules such as H_2O and CO_2: these are known as the Advanced Oxidation Processes. Heterogeneous photocatalysis is one of the most investigated degradation processes, with the greatest performance, respect to the environment, and this efficiency has been proven in the degradation of pollutants [22, 23, 32].

Several approaches to manufacturing stable chemically composite as pollutants degradation are progressing rapidly [43]. Usually, a single material exhibits catalytic activity for a specific dye. However, this particular material may not permanently be able to provide a similar catalytic activity for different dyes. Therefore, research is being done to develop composites [10, 26, 48] or nanocomposite [1, 31, 41] which can give superior performance in a multi-dye system. This chapter presents an outline of the application of composite materials for dye photocatalytic degradation and presents a different synthesis technique, as well as the progress of the photocatalytic removal process vs conventional methods.

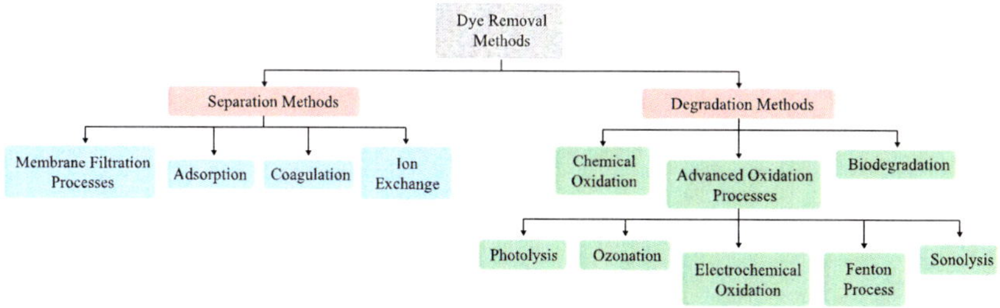

Fig. 1 The dye removal techniques

2 Dyes Degradation

Various treatment processes have been employed to eliminate synthetic dyes from the solution. The process of dyes elimination can be arranged into two classes:

- Separation: by applying a physical or physical-chemical process.
- Degradation: thanks to the chemical and biological reaction process.

The separation processes are the most used technique, however, the main disadvantage is the disposal of dye-containing sludge. During the degradation procedure, the complex dye compounds undergo a series of degradation (Fig. 1).

3 Advanced Oxidation Process

The process basic principle is focused on the generation of highly reactive oxidants, primarily hydroxyl radicals (OH), the radical (OH) with a redox potential OH/H_2O of 2.8 V/SHE, which is classified as one of the stronger oxidizing agents after fluorine [6].

These radical attacks the organic pollutants with high reaction rate with a capacity to destroy most organometallic and organic pollutants, by the conversion to CO_2, H_2O, and organic ions, the process achieved generally through the following three different ways:

- Redox reaction $OH + RX \rightarrow RX^+ + OH^-$
- Dehydrogenation: consist of abstraction of H atom $OH + RH \rightarrow R + H_2O$
- Hydroxylation: consist of an electrophonic addition to π system $OH + PhOH \rightarrow PhXOH$

These reactions create organic radicals, which initiate a radical chain that includes reactions with oxygen (formation of peroxy radicals), and the produced reaction intermediates, undergo additional oxidation reactions with the created oxidizing agents (OH^-, HO_2, H_2O_2) until the organic contaminants are fully mineralized.

Electrochemical advanced oxidation processes use electrolytically formed hydroxyl radicals to mineralize organic contaminants; this process is separated into two classes: Direct and indirect processes [36].

Direct process: hydroxyl radicals are formed on the anode surface as a result of direct oxidation of water according to:

$$M + H_2O \rightarrow M(OH) + H^+ + e^-$$

$$2O \rightarrow M(OH) + H^+ + e^-$$

The catalytic behavior of the anode material is determined by its composition. The generation of hydroxyl radicals in this method does not necessitate the addition of any additional reagent. Yao et al. [53] prepared nanocomposite electrodes of PbO_2-ZrO_2 by pulse electrodeposition and successfully removed 10% of methylene blue.

Indirect process: The Fenton chemistry is used to generate hydroxyl radicals, which can be done in situ (electro-Fenton) or by adding external reagents like H_2O_2 or ferrous ion.

3.1 Fenton Process

Fe^{2+} and H_2O_2 are formed simultaneously in the electro Fenton process by cathodic reduction of Fe^{3+} and O_2, respectively. A small amount of Fe^{2+} salt is added first, which can react with electrochemically produced H_2O_2 to produce Fe^{3+}.

The following equation is often used to explain the mechanism of the photo-Fenton reaction.

$$Fe^{2+} + H_2O_2 \rightarrow Fe^{3+} + OH^- + \dot{O}H$$

$$Fe^{3+} + H_2O_2 \rightarrow Fe^{2+} + HOO^{\cdot} + H^+$$

$$Fe^{3+} + H_2O \xrightarrow{h\upsilon} Fe^{2+} + \dot{O}H + H^+$$

$$H_2O_2 \xrightarrow{h\upsilon} 2\,^{\cdot}OH$$

$$^{\cdot}OH + \text{organic compound} \rightarrow \text{oxidized products}$$

The rate of degradation of the photo Fenton reaction increases in the presence of UV light due to the $Fe^{(3+)}$ photoreduction, which produces a current. $Fe^{(2+)}$ ions are formed by the OH radical, and $Fe^{(2+)}$ can react with H_2O_2.

Photolysis of H_2O_2 can produce $OH^{\bullet}$ radical directly, this synergetic effect causes an enhancement of the catalytic activity for photo-Fenton, at the industrial scale the heterogeneous Fenton process is more advantaged compared to homogeneous Fenton, they lead to scale up and catalyst reuse, They reduce the residual concentration of an iron ion in the treated solution, and in a heterogeneous phase, Fe ions may be loaded onto several substrates to create a catalytic composite.

3.2 Nanocomposite in Fenton-Based Process

Many issues with this technique, such as stabilization, recycling, and isolation, can be solved by immobilizing the catalyst on stable supports. Both organic and inorganic materials, such as activated carbon, have been used as supports for Fenton-like reactions [33], fiber [42], clay [13], and carbon nanotube.

A different form of iron oxides and iron hydroxides such as Fe_3O_4, γ-Fe_2O_3, α-Fe_2O_3, α-FeOOH, and γ-FeOOH have been investigated as catalysts for Fenton like processes.

The exploration of nanocarbon materials, is rapidly evolving in the catalysis field as an opportunity to improve the catalyst propriety, the nanocarbon supported nanocatalyst such as carbon nanotube (nanocarbon/metal oxide) is synthesized and used for the degradation of various dyes.

Carbon nanotube (CNT) supported iron catalysts (Fe_2O_3/CNT) and Fe double-walled CNT mixed metal oxide, have been synthesized and tested for their ability to degrade (RhB), Fe_2O_3/MWCNT (multiwalled carbon nanotube) materials for Methyl orange degradation were tested [51]. The effectiveness has improved compared to the unsupported Fe_3O_4, for example in the case of Fe_3O_4/MWCNT the achieved degradation efficiency was two times greater than Fe_2O_3 for acid orange II decomposition [51].

Graphene oxide (GO) and Iron oxide have been emerged as favorable composites, the graphene and GO used due to their outstanding electron transfer capacity, good stability, and high surface area.

Graphene oxide supported Fe_2O_3 prepared by Guo et al. [17], the degradation of RhB was measured, in the presence of H_2O_2, depending on the amount of GO (5–20%) the effect on the degradation have been reported to fluctuate between 2 and 5 time more compared to unsupported Fe_2O_3, However, when the loading quantity surpasses 25% the removal affected negatively. Palanivel et al. [38] prepared GO-doped $NiFe_2O_4$ nanocomposite following the sol-gel and hydrothermal method as a photo-Fenton catalyst for organic RhB dye degradation.

Another composite class based on Clay such as laponite and bentonite supported iron oxide was prepared for dye degradation in the Fenton process [45].

As cathode material in electro Fenton application, the nanocomposite is an emerging trend, nanostructure such as Fe_2O_3/CNT used as an electrode for the degradation of RhB, as a cathode material was able to produce OH radicals and degrade RhB in a wide range of pH, many nanocomposites employed as cathode have been studied such Fe_2O_3/activated carbon, Fe_2O_3/CNT/PTFE, carbon nanotube immobilized graphite cathode combined with anodes such as Pt sheet or wire.

Alimard [2] prepared Neodymium-cesium (Nd–Ce) doped Fe_3O_4-chitosan combining co-precipitation and a crosslinking process and evaluated on the degradation of Red 81 dye.

Tertiary composite such as ZnO deposition on Fe_3O_4/rGO [28], the nanocomposite, which was made by hydrothermally depositing of ZnO on Fe_3O_4/GO prepared via a solvothermal process, showed superior catalytic activity for the degradation of Azo dyes when compared to ZnO, Fe_3O_4, and a ZnO/Fe_3O_4 bicomposite. However, using a green one-pot annealing process, whit N-doped carbon/CuO-Fe_2O_3 (NC-CuFe) was made, height MB dyes degradation was observed [40].

Using co-precipitation method combined graft co-polymerization, Cellulose based hydrogel coated Fe_3O_4 prepared by Zhou et al. [57] and used as Fenton-like catalyst for acid red 73 degradations, in H_2O_2 activation the composite showed high efficiency.

3.3 Photocatalysis: State of the Art and Basic Principles

Photocatalysis is an emerging technology based on the characteristic of the semiconductor, the titania dioxide (TiO_2) is the most famous semiconductor used as a catalyst, the photoactivity of TiO_2 was investigated in the late 1930s and their ability to perform as a photosensitize was revealed, when it was incorporated as a pigment in the building found a bleaching effect under solar irradiation and in the presence of oxygen, The work of Fujishima has had a major impact on the interest in TiO_2 as photoactive material (1970) [15] and Nozik [37], since their application is enormously boosted and focused recently on the development of more efficient and sophisticated photocatalytic materials (Fig. 2).

One of the targeted results for a successful photocatalyst is its aptitude to harvest light absorption to bring it to a higher energy level, When irradiation coincides or surpasses the bandgap energy of the semiconductor which is 3.2 eV for anatase TiO_2, the electrons are ejected to the conduction band which leads a formation of a hole, the holes reach the catalyst surface and react with the adsorbed hydroxyl group to form OH radical, that leave the surface as free radicals (OH). With the discovery of the capacity of the anatase nanocrystalline TiO_2 to perform this function under UV illumination, these results were followed by many suggestions for doped TiO_2 and other alternative materials for example composite nanofibers systems containing cellulose acetate, multiwall carbon nanotubes to increase the efficiency that could be extended into the visible range.

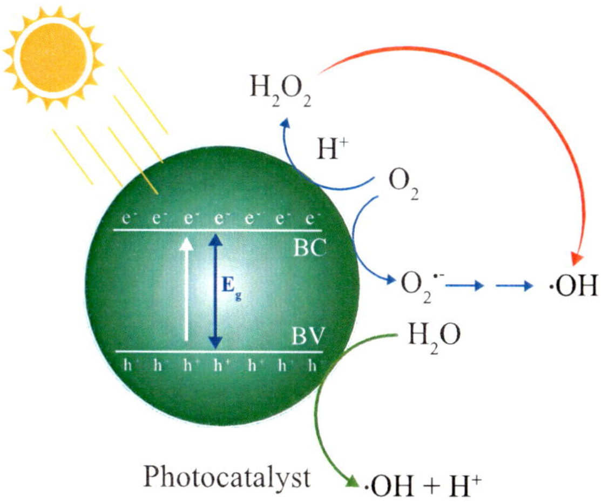

Fig. 2 Representation of the photocatalytic degradation mechanism

3.4 Nanocomposite Preparation for Photocatalytic Dyes Degradation

Nanocomposite material for photodegradation efficiency improvement in the irradiation of visible light is dependent on the combining of photocatalyst material or support with a narrower bandgap metal-semiconductor, these include mixed metal oxide, nonporous, polymers, and carbon-based nanocomposite.

Recently many nanocomposites have been prepared and exhibited a high photocatalytic activity mainly to overcome some technical limitations such as the recycling and the recovering of the nanocatalyst powders and the risk of their release into the environment; many efforts have been achieved on the immobilization and incorporation.

Three main steps are necessary to accomplish the successful preparation of a photocatalyst nanocomposite

- the choice and synthesis of an effective photocatalyst
- surface functionalization in order solid linking with the support
- incorporation into a suitable matrix depending on the final utilization.

a. **Carbon nanofibre: Amorphous carbon nitride**

Carbon nitrides material is a combination of carbon and nitrogen in a ratio of 3–4, can easily be obtained at low cost from a nitrogen-rich material, their proprieties can be designed according to the synthesis strategies, precursors, and polymerization condition, nitrogen content depends primarily on the content in the used precursor.

Nitride is classified as semiconductors with a narrower bandgap compared to metal oxide semiconductors, they are superior to oxide semiconductors [12]. The increased visible light capture of nitride semiconductors is due to the nitrogen (N) having a lower electronegativity than oxygen (O), and the N2p orbital having higher potential energy than the O2p orbital. The graphitic carbon nitride (g-C_3N_4) semiconductor has many allotrope phases, including diamond-like, -C_3N_4 cubic, pseudocubic, and polymer-like g-C_3N_4. At atmospheric conditions, the amorphous graphitic phase C_3N_4 has the most stable allotrope type.

g-C_3N_4 with a bandgap of 2.7 eV is considered as an indirect semiconductor, they facilitate the absorption of visible light. Graphitic C_3N_4 and ferrite composite ($ZnFe_2O_4$-gC_3N_4) [52] prepared via facile sol-gel followed by calcination method, The composite shows the photo Fenton like reaction and fastest degradation capability in the presence of H_2O_2 in all visible light sources as used for sunlight-driven photocatalytic organic pollutant removal.

The electronegativity of g-C_3N_4 and $ZnFe_2O_4$ is around 4.7 and 5.8 eV, respectively, their bandgap energy is between 2.75 and 1.73 eV which matches with pristine g-C_3N_4 and $ZnFe_2O_4$ the value of the estimated E_{cb} and $E_{\mathbf{vb}}$ of g-C_3N_4 is (−1.14 eV) and (1.60e). The mechanism of dyes degradation in the presence of $ZnFe_2O_4$-gC_3N_4 can be explained as follows:

When a photon interacts with the photocatalyst's surface, on both catalysts the electrons are dislocated from the valance band to the conduction band. The excited conduction band electron in the g-C_3N_4 migrated to the $ZnFe_2O_4$ conduction band due to low potential energy, though the holes migrated to the g-C_3N_4 valance band. Under light irradiation, the dye molecule can be degraded in two ways: the O_2 is reduced to O_2* by the conduction band electron during the transmittance at the g-C_3N_4 and the O_2 is reduced to O_2* by the conduction band electron during the transmittance at the g-C_3N_4. The following is the breakdown mechanism:

$$ZnFe_2/gC_3N_4 + h\nu \rightarrow ZnFe_2O_4\left(e^-/h^+\right)$$

$$gC_3N_4 + O_2 \rightarrow O_2\text{-}$$

$$O_2\text{-} + H_2O \rightarrow H_2O_2$$

$$O_2\text{-} + H_2O \rightarrow OH^- + OH^*$$

$$OH^* + \text{dye} \rightarrow CO_2 + H_2O$$

The electron on the conduction band of $ZnFe_2O_4$ interacts with H_2O_2 to create OH* radicals. One of the advantages of magnetic particle is their aptitude to use the magnetic field to recover the nanocomposite for the recycling process after each reaction as example graphitic carbon nitride/copper-iron oxide (CN@CIO) prepared via in situ formation of g-C_3N_4 onto copper iron oxide.

The forming of chemical bonds during calcination as a result of metal oxide intercalation produced several layer gaps in the g-C_3N_4 structure.

b. Clays-based nanocomposite

Clay refers to a class of layered materials made up of silicates or minerals with metal oxides and organic matter, Clay minerals are promising support materials because they exhibit high specific surface areas, large pore sizes, chemical stability, strong mechanical properties, and they are commercially available in large quantities. Clay minerals, such as kaolinite which a layered silicate and aluminum oxide/hydroxide, with the formula $Al_2Si_2O_5(OH)_4$.

The kaolinite basic structure is constituted by one tetrahedral (Si-O) and one octahedral (Al-O) sheet, which is held together by hydrogen bonding between the silicate and aluminum layers. The 1:1 clay (Kaoline) formed by One tetrahedral silica layer is lined on top of octahedral gibbsite or brucite sheet, however, the montmorillonite has two tetrahedral silica and one octahedral gibbsite (2:1)

Clays materials tend to boost the photocatalytic activity of TiO_2 nanoparticles, this a consequences of their elevated surface area, adsorption capacity, and large porosity with the availability of the active site TiO_2/clay, the presence of interlayer cations catch electrons and make the holes free for oxidation which makes charge recombination of TiO_2/Clay decrease, for the preparation of nanocomposite, impregnation methods are highly preferred (clay surface or it's layered with TiO_2). The sol-gel method combined with different process dryings, such as air or supercritical drying, ethanol extraction, the hydrothermal method preferred to the realization of highly crystalline TiO_2 (anatase 40–60°A). However, Bentonite mostly constituted of montmorillonite, acid-catalyzed sol-gel used for TiO_2 based Bentonite nanocomposite.

c. Graphene (porous, graphene oxide nanosheet)

Graphene is a carbon isotope with an excellent physiochemical property added to its various applications, graphene is an interesting candidate for photocatalysis application as graphene-based composite, and a variety of semiconductor photocatalyst has been used for the preparation of graphene-based composites employing a range of synthesis process such as hydrothermal, solvothermal method.

TiO_2/graphene composite prepared by one-step hydrothermal, this consist on the loading of the semiconductor or the metal precursors on the GO sheets and then rGO, the nanocomposite showed an important improvement in the reaction rate, and high-performance photocatalyst on the degradation of methylene blue (MB), and the activities of their activity found to be more active than the unsupported catalyst.

The composite $ZnFe_2O_4$-graphene prepared by hydrothermal in ethanol aqueous solution, same as $Bi_2WO_6TiO_2$/GO nanocomposite which enhances the MB degradation under UV by a factor of 1.4 compared to TiO_2, however on methyl orange the efficiency is very low even though the factor is improved depending on the graphene content.

d. **Polymer-based nanocomposite**

Such as biopolymer hydrogel, polyaniline, cellulose, gar gum, chitosan grafted polyaniline or Synthetic polymers such polyethylene, polypyrrole, polyaniline, polypropylene (PP), polylactic acid (PLA), poly (3-hexylthiophene), polyvinyl alcohol has been used to prepare a nanocomposite as materials for photocatalytic application for dyes degradation, polymers matrices is an attractive host thanks to their surface chemistry and controllable pore size, the polymer with comfort and excellent mechanical propriety, the choice of the polymer to integrate as polymeric nanocomposite for dyes degradation may depend on the thermomechanical behavior, biocompatibility, surface proprieties, chemical, and thermal stability, electronic or electric propriety and many other characteristics could be considered, different technologies have been investigated for preparing polymers based nanocomposite.

e. **In situ synthesis**

In situ synthesis is widely applied in different type, the catalyst could be applied sol-gel approach has been employed for catalyst polymer, such as polyimide TiO_2 hybrid films from soluble polyimides, the TiO_2 dropped into polyimide solution with vigorous stirring with the amount of water to assist the hydrolytic reaction of TiO_2 followed by filtration and cast heated a 250 °C to transformer the gel into the hybrid film, another approach is the polymerization of the monomer containing catalyst this method allows to synthesize the nanocomposite with tailored physical properties, ZnO was traded with methacryloxypropyl trimethoxysilane and disperse in methyl methacrylate [11].

f. **Template synthesis**

Template synthesis is a simple procedure which has made the fabrication of sophisticated nonmaterial's relatively easy, nanofibers, microporous membrane an many other solid could be obtained by this method, the synthesis procedure achieved thanks to a polymer template containing cylindrical pore of uniform pore diameter, the pores filling process could be achieved via many processes like sol-gel, electrodeposition, chemical vapor deposition.

g. **Electrospinning**

Electrospinning is a method of producing polymer, ceramic, metallic, and composite fibers from a solvent dispersion using a syringe pump to move a suspension to the needle edge, a high voltage applied to the metallic needle, and a syringe., metal oxide and catalyst could be included in the electrospun polymer fibers, Nylon-6/TiO_2 nanocomposite prepared with TiO_2 solution and 20%wt Nylon in formic acid and acetic acid the electrospinning process was achieved with a voltage of 18 kV, the nanocomposite an ability to be recycled offered a possibility to degrade a methylene bleu.

Conventional solvent casting technique has been used to prepare a polymeric film with doping oxides, the key elements for this technique are that the polymer must be soluble in a volatile solvent or water, and a solid solution, nanosized TiO_2 photocatalyst were immobilized onto polyhydroxy butyrate films prepared by dissolving it in chloroform and participated in cold methanol, under a solar light 96% of dye decolorized after an hour, the cellulose/TiO_2 nanocomposite reported.

In the presence of supercritical CO_2, the molecular chains of cellulose were affected, and TiO_2 was facilitated to enter and impregnate into the structure of cellulose fibers through the forming of hydrogen bonds with abundant hydroxyl groups of cellulose, resulting in a significant increase in thermal stability.

Polystyrene film-based photocatalytic film was prepared using solvent casting method incorporating rGo graphitic carbon along with TiO_2 as catalyst were prepared 94% of decolorization of complex phthalocyanine dye wastewater.

h. **Activated carbon**

Activated carbons (ACs) with different porosities are an excellent candidate for photocatalytic applications, as environmentally stable and low-cost materials have been used as hot and support for metal oxide and photocatalytic catalysts, often obtained by dip hydrothermal process.

The loading of the catalyst on the activated carbon has been achieved in the reported work in a different way, such by impregnation (Ag/AC, ZnO/AC), co-precipitation (Ag-ZnO/AC), or by consecutive precipitation TiO_2 prepared by sol-gel and supported on spherical AC to prepare TiO_2/AC hybrid photocatalysts efficiency removal of ZnO/AC was more than two times compared to ZnO [49]. The role of porous activated carbon used as a catalyst for carbon/TiO_2 composite in photodegradation of phenol was investigated and the presence of a synergistic impact of carbon on photodegradation operation was demonstrated. SnO_2/corn cob activated carbon showed as best photocatalytic toward the degradation of MB when compared to SnO_2 alone a synergic effect of both adsorption and photocatalysis have been shown with a reduced bandgap.

ZnO nanorod assembled on AC fibers via sol-gel and hydrothermal process exhibited a high degradation rate of MB dye compared to ZnO powder, CdS nanoflower loaded AC by the mixed solvothermal route and tested for RhB degradation, the main advantage of activated carbon they allow a high adsorption capacity on the surface catalyst and assist charge efficient separation, which eventually enhances the photocatalytic performance, as a consequence of the synergistic effect of adsorption and photocatalysis.

Sol impregnation hydrothermal process has been applied to prepare a bifunctional composite supported on activated carbon such as Bi_2O_3/TiO_2, on methylene orange degradation the composite showed high efficiency.

i. **Magnetic nanocomposite**

Magnetic nanocomposite for the photocatalytic application has been emerged as a potential alternative to assist the separation process and easily recoverable catalyst during the liquid phase reaction, the isolation of magnetic catalyst from the reaction medium based on applying a simple magnetic field, especially for small catalysis particles.

Various magnetic photoactive catalysts have been prepared with different processes, Single-phase catalysts, nanocomposites, multifunctional metal-organic matrix binary, ternary, and core-shell materials are only a few examples.

Single-phase magnetic materials such as magnetite Fe_3O_4, magnetite (γ-Fe_3O_4 and β-Fe_3O_4), and hematite (α-Fe_3O_4). The combining of magnetic photocatalyst oxide with another oxide is a new technique for preparing highly photoactive materials, and the combination of two distinct oxides will also reduce charge recombination and increase catalytic activity. The single component has a low potential for absorption in the visible range, which contributes to poor catalytic efficiency.

A lot of composite was prepared based on the hydrothermal process and evaluated on RhB like $MoS_2/CoFe_2O_4$, Fe_3O_4/Ag, $Bi_{25}FeO_{40}/Fe_3O_4/Fe_2O_3$, the catalytic activity remained stable after many utilizations. ZnO (gap energy of 3.37 eV) compared to Anatase TiO_2 (3.20 eV) and Rutile 3.03 eV which is rarely used.

As a magnetic nanoparticle the Fe_2O_4, which made by co-precipitation of $FeCl_2$, $FeCl_3$ in an aqueous solution under alkaline condition, the magnetic particle integrated with a catalyst in a various process such as sol-gel by mixing the magnetic particles with a precursor such as tetrabutyl titanate dissolved previously in ethanol to obtain a Fe_2O_4/TiO_2 [24, 29].

j. **Multimetal oxide Semiconductor/metal oxide heterostructures**

To boost the visible light photocatalytic activity, semiconductors have been coupled with other semiconductors or metal oxide heterostructures with a narrower bandgap.

To be photocatalytically active under visible light, a catalyst must have a narrower bandgap and lower charge carrier recombination, all semiconductors suffer from charge carriers recombination either in the bulk or on the surface many approaches have been applied to exhibit a high quantum yield, firstly by modifying the catalyst by substitution of a metal cation or nonmetal anion, or by co-doping with compensating cation-anion pairs, another alternative is by exploiting the structural engineering of nanostructure like synthesizing and the more reactive (001) anatase surface rather than (101). The recent advanced result showed that the heterostructure and metal oxide semiconductor combination is a successful approach to adjust the structure to give a visible light absorption and improved the photocatalytic activity compared to pure oxide, composite like BiOBr-$ZnFe_2O_4$, $BiVO_4$-WO_3, CdS-TiO_2, WO_3-TiO_2, Bi_2WO_6, combining of ZnO with Ag, CdO, or in a ternary form such as ZnO/Ag/Mn_2O_3, MoO_3/TiO_2, V_2O_5/TiO_2, for example, in the case of (RhB) degradation, a metal oxide such as Bi_2WO_6/SnS showed an efficiency of 66% [46], however, Bi_2O_3-CeO_2ZnO [21], ZrO_2/PbO_2 [53] using single-step hydrothermal method showed en efficiency of 93% $BiVO_4$/$Bi_2O_2CO_3$, CdO-ZnO (97%), CeO_2-Y_2O_3 (98%) on the other hand NiO-CdO-ZnO [34] prepared by homogeneous co-precipitation method achieved a degradation efficiency of 99 and 98% for methylene blue.

Some metals have been used on the degradation of MB such as CdO-NiO, Fe_2O_3-CuO-ZnO, CdO-NiO ZnO, CdO-ZnO-MgO (78–91%).

Recently a newly emerging strategy based on an environmentally friendly, economic, and simple process to prepare a nanocomposite based on the green synthesis, Au/TiO_2 composite prepared via a green process, in the presence of Degussa P-25 TiO_2 $HAuCl_4$ reduced with, *Cinnamoum tamala* leaves the aqueous extract, the degradation activity of the composite on methyl orange found to be 2.5 times more compared TiO_2 only [35], CuO/ZnO nanocomposites prepared by using $Cu(CH_3COO)_2 \cdot H_2O$ and $Zn(CH_3COO)_2 \cdot 2H_2O$ precursors and biomass filtrate (*Penicillium corylophilum*). The ratio of CuO (CuO/ZnO) give a noticeable decrease in the optical band gap 6.32 eV (80%CuO) to 5.22 eV (20%CuO) [14] (Table 1).

4 Conclusion

Important progress has been made in the use of nanocomposites in dye degradation in recent years. These nanocomposites offer better performance compared to micro composite materials. There are several methods for making nanocomposite materials, each with its characteristics. Today, the economic challenges for the development of these materials are multiple and already affect many fields of application such as construction, transport, food packaging, sporting goods, textiles, etc.

Currently, advanced oxidation systems may be an alternative to traditional approaches for the degradation of organic contaminants.

Table 1 Nanocomposite in Fenton process for dyes degradation

Composite	Preparation method	Dyes degradation	References
rGO supported $NiFe_2O_4$/g-C_3N_4	Sol-gel and hydrothermal	Photo-Fenton degradation of RhB	Palanivel et al. [38]
Nickel on hydroxyapatite-encapsulated with Cobalt Ferrite (Ni/HAP/$CoFe_2O_4$)	Sol-gel precipitation	Fenton-MB and methyl orange	Das et al. [9]
ZnO/FeVO4	Coprecipitation	Photo-Fenton like process	Rahimpour et al. [39]
Iron gel-chitosan/clay (montmorillonite nanosheets)(Fe-CS/MMTNS)	Self-assembled	Photo-Fenton MB dye	Palanivel et al. [38]
$MnFe_2O_4$/rGO	One-pot hydrothermal method	Photo-Fentonmethylene blue (MB)	Wei et al. [50]
Co Fe_2O_4/rGO	Thermal decomposition	Fenton-like reaction – methyl orange (MO) – Rh B	Hassani et al. [20]
$MnFe_2O_4$@ SnS_2	Hydrothermal	Photo-Fentonmethylene blue (MB)	Zhao et al. [56]
MnO_2/NiO nanocomposite	Hydrothermal method	Photo-Fenton-like degradation of orange II	Zhang et al. [55]
Fe_3O_4@cellulose	Co-precipitation	Fenton reaction methylene blue (MB)	Lu et al. [30]
Fe_3O_4@cellulosaerogel	Hydrothermal	Fenton-like degradation of RhB	Jiao et al. [25]
Fe_3O_4/MWCNT	Solvothermal	Heterogeneous Fenton methyl orange	Xu et al. [51]
Fe_2O_3/CNT	Chemical synthesis	Heterogeneous E-Fenton RhB	Yao et al. [54]
$ZnFe_2O_4$@g-C_3N_4	Co-precipitation	Fenton-like photodegradation RhB	Borthakur and Saikia [4]
Fe_3O_4/TiO_2/C nanocomposites	Precipitating hydrothermal	MB dye Fenton-like system	Liu et al. [29]
(Fe_3O_4 NPs) and molybdenum disulfide microspheres (MoS_2)	Hydrothermal	Fenton-like RhB	Han et al. [19]

Acknowledgements The authors acknowledge the financial support from the Directorate-General for Scientific Research and Technological Development (DGRSDT, Algeria).

References

1. Albiter E et al (2020) Synthesis, characterization, and photocatalytic performance of ZnO–graphene nanocomposites: a review. J Compos Sci 5(1):4
2. Alimard P (2019) Fabrication and kinetic study of Nd-Ce doped Fe_3O_4-chitosan nanocomposite as catalyst in Fenton dye degradation. Polyhedron 171:98–107. http://www.sciencedirect.com/science/article/pii/S027753871930470X
3. Bafana A, Devi SS, Chakrabarti T (2011) Azo dyes: past, present and the future. Environ Rev 19(NA):350–371. https://doi.org/10.1139/a11-018
4. Borthakur S, Saikia L (2019) $ZnFe_2O_4$@g-C_3N_4 nanocomposites: an efficient catalyst for Fenton-like photodegradation of environmentally pollutant rhodamine B. J Environ Chem Eng 7(2):103035. http://www.sciencedirect.com/science/article/pii/S2213343719301587
5. Boyles C, Sobeck SJS (2020) Photostability of organic red food dyes. Food Chem 315:
6. Buthiyappan A, Raman AAA, Daud WMAW (2016) Development of an advanced chemical oxidation wastewater treatment system for the Batik Industry in Malaysia. RSC Adv 6(30):25222–25241. https://doi.org/10.1039/C5RA26775G
7. Carmen Z, Daniel S (2012) Textile organic dyes—characteristics, polluting effects and separation/elimination procedures from industrial effluents—a critical overview. In: Organic pollutants ten years after the Stockholm convention—environmental and analytical update. IntechOpen, p 55
8. Coultate T, Blackburn RS (2018) Food colorants: their past, present and future. Color Technol 134(3):165–186
9. Das KCh, Das B, Dhar SS (2020) Effective catalytic degradation of organic dyes by nickel supported on hydroxyapatite-encapsulated cobalt ferrite (Ni/HAP/$CoFe_2O_4$) magnetic novel nanocomposite. Water Air Soil Pollut 231(2):43. https://doi.org/10.1007/s11270-020-4409-1
10. Derikvand L, Tahmasebi N (2021) Synthesis and photocatalytic performance of Bi_2WO_6/BiOX (X = Cl, Br, I) composites for RhB degradation under visible light. Korean J Chem Eng 38(1):163–169
11. Di Mauro A et al (2017) Novel synthesis of ZnO/PMMA nanocomposites for photocatalytic applications. Sci Rep 7(1):40895. https://doi.org/10.1038/srep40895
12. Dong G, Zhang Y, Pan Q, Qiu J (2014) A fantastic graphitic carbon nitride (g-C_3N_4) material: electronic structure, photocatalytic, and photoelectronic properties. J Photochem Photobiol C: Photochem Rev 20:33–50. http://www.sciencedirect.com/science/article/pii/S1389556714000203
13. Feng J, Xijun H, Yue PL (2004) Novel bentonite clay-based Fe—nanocomposite as a heterogeneous catalyst for photo-Fenton discoloration and mineralization of orange II. Environ Sci Technol 38(1):269–275
14. Fouda A et al (2020) Heliyon optimization of green biosynthesized visible light active CuO/ZnO Nano-photocatalysts for the degradation of organic methylene blue dye. Heliyon 6(September): https://doi.org/10.1016/j.heliyon.2020.e04896
15. Fujishima A, Honda K (1972) Electrochemical photolysis of water at a semiconductor electrode. Nature 238(5358):37–38. https://doi.org/10.1038/238037a0
16. Guerra E, Llompart M, Garcia-Jares C (2018) Analysis of dyes in cosmetics: challenges and recent developments. Cosmetics 5(3)
17. Guo S, Zhang G, Guo Y, Yu JC (2013) Graphene oxide–Fe_2O_3 hybrid material as highly efficient heterogeneous catalyst for degradation of organic contaminants. Carbon 60:437–444. http://www.sciencedirect.com/science/article/pii/S0008622313003540
18. Gürses A, Açıkyıldız M, Güneş K, Sadi Gürses M (2016) Dyes and pigments: their structure and properties. Springer, Cham, pp 13–29
19. Han C, Huang G, Zhu D, Hu K (2017) Facile synthesis of MoS_2/Fe_3O_4 nanocomposite with excellent photo-Fenton-like catalytic performance. Mater Chem Phys 200:16–22. http://www.sciencedirect.com/science/article/pii/S0254058417305849

20. Hassani A et al (2018) Heterogeneous sono-Fenton-like process using magnetic cobalt ferrite-reduced graphene oxide ($CoFe_2O_4$-RGO) nanocomposite for the removal of organic dyes from aqueous solution. Ultrason Sonochem 40:841–852. http://www.sciencedirect.com/science/article/pii/S1350417717303802
21. Hezam A et al (2017) Synthesis of heterostructured Bi_2O_3–CeO_2–ZnO photocatalyst with enhanced sunlight photocatalytic activity. Ceram Int 43(6):5292–5301. http://www.sciencedirect.com/science/article/pii/S027288421730072X
22. Huang CW et al (2020) Metal-organic frameworks: preparation and applications in highly efficient heterogeneous photocatalysis. Sustain Energy Fuels 504–521
23. Iervolino G, Zammit I, Vaiano V, Rizzo L (2020) Limitations and prospects for wastewater treatment by UV and Visible-light-active heterogeneous photocatalysis: a critical review. Top Curr Chem 378(1):7
24. Jacinto MJ, Ferreira LF, Silva VC (2020) Magnetic materials for photocatalytic applications—a review. J Sol-Gel Sci Technol 96(1). http://dx.doi.org/10.1007/s10971-020-05333-9
25. Jiao Y et al (2018) Facile hydrothermal synthesis of Fe_3O_4@cellulose aerogel nanocomposite and its application in Fenton-like degradation of rhodamine B. Carbohydr Polym 189: 371–378. http://www.sciencedirect.com/science/article/pii/S0144861718301760
26. Khurana I et al (2017) Removal of dyes using graphene-based composites: a review. Water Air Soil Pollut 228(5):1–17
27. Kumar A, Konar A (2011) Dyeing of textiles with natural dyes. In: Natural dyes. IntechOpen, pp 30–56
28. Lei Y et al (2020) Low-temperature preparation of magnetically separable Fe_3O_4@ZnO-RGO for high-performance removal of methylene blue in visible light. J Alloys Compd 821:153366. http://www.sciencedirect.com/science/article/pii/S0925838819346122
29. Liu X et al (2016) Preparation of Fe_3O_4/TiO_2/C nanocomposites and their application in Fenton-like catalysis for dye decoloration. Catalysts 6(9)
30. Lu Q et al (2019) In situ synthesis of a stable Fe_3O_4@cellulose nanocomposite for efficient catalytic degradation of methylene blue. Nanomaterials 9(2). https://www.mdpi.com/2079-4991/9/2/275
31. Lum PT, Foo KY, Zakaria NA, Palaniandy P (2019) Ash based nanocomposites for photocatalytic degradation of textile dye pollutants: a review. Mater Chem Phys 122405. https://doi.org/10.1016/j.matchemphys.2019.122405
32. Martínez C et al (2011) Aqueous degradation of diclofenac by heterogeneous photocatalysis using nanostructured materials. Appl Catal B 107(1–2):110–118
33. Messele SA et al (2012) Phenol degradation by heterogeneous Fenton-like reaction using Fe supported over activated carbon. Procedia Eng (Elsevier Ltd.) 1373–1377
34. Munawar T et al (2020) Multi metal oxide NiO-CdO-ZnO nanocomposite–synthesis, structural, optical, electrical properties and enhanced sunlight driven photocatalytic activity. Ceram Int 46(2):2421–2437. http://www.sciencedirect.com/science/article/pii/S0272884219327749
35. Naik GK, Mishra PM, Parida K (2013) Green synthesis of Au/TiO_2 for effective dye degradation in aqueous system. Chem Eng J 229:492–497. https://doi.org/10.1016/j.cej.2013.06.053
36. Nidheesh PV, Zhou M, Oturan MA (2018) An overview on the removal of synthetic dyes from water by electrochemical advanced oxidation processes. Chemosphere 197:210–227. http://www.sciencedirect.com/science/article/pii/S0045653517321732
37. Nozik AJ (1977) Electrode materials for photoelectrochemical devices. J Cryst Growth 39(1):200–209. http://www.sciencedirect.com/science/article/pii/002202487790166X
38. Palanivel B, Alagiri M (2020) Construction of RGO supported integrative $NiFe_2O_4$/g-C_3N_4 nanocomposite: role of charge transfer for boosting the OH. Radical production to enhance the photo-Fenton degradation. ChemistrySelect 5(31):9765–9775. https://chemistry-europe.onlinelibrary.wiley.com/doi/abs/10.1002/slct.202002519
39. Rahimpour R, Chaibakhsh N, Zanjanchi MA, Moradi-Shoeili Z (2020) Fabrication of ZnO/$FeVO_4$ heterojunction nanocomposite with high catalytic activity in photo-Fenton-like process. J Alloys Compd 817:152702. http://www.sciencedirect.com/science/article/pii/S0925838819339489

40. Ren B et al (2019) A grape-like N-doped carbon/CuO-Fe_2O_3 nanocomposite as a highly active heterogeneous Fenton-like catalyst in methylene blue degradation. J Clean Prod 240:118143. http://www.sciencedirect.com/science/article/pii/S0959652619330136
41. Riaz U, Ashraf SM, Kashyap J (2015) Role of conducting polymers in enhancing TiO_2-based photocatalytic dye degradation: a short review. Polym Plast Technol Eng 54(17):1850–1870
42. Seyed Dorraji MS et al (2015) Fenton-like catalytic activity of wet-spun chitosan hollow fibers loaded with Fe_3O_4 nanoparticles: batch and continuous flow investigations. J Mol Catal A: Chem 398:353–357
43. Shindhal T et al (2021) A critical review on advances in the practices and perspectives for the treatment of dye industry wastewater. Bioengineered 12(1):70–87
44. Šuleková M et al (2017) Organic colouring agents in the pharmaceutical industry. Folia Veterinaria 61(3):32–46
45. Sze Nga Sum O, Feng J, Hu X, Yue PL (2004) Pillared laponite clay-based fe nanocomposites as heterogeneous catalysts for photo-Fenton degradation of acid black 1. Chem Eng Sci 59(22):5269–5275. http://www.sciencedirect.com/science/article/pii/S000925090400692X
46. Tang R et al (2016) Enhanced photocatalytic performance in Bi_2WO_6/SnS heterostructures: facile synthesis, influencing factors and mechanism of the photocatalytic process. J Colloid Interface Sci 466:388–399. http://www.sciencedirect.com/science/article/pii/S0021979715304380
47. Tkaczyk A, Mitrowska K, Posyniak A (2020) synthetic organic dyes as contaminants of the aquatic environment and their implications for ecosystems: a review. Sci Total Environ 717:
48. Varadavenkatesan T et al (2021) Synthesis, biological and environmental applications of hydroxyapatite and its composites with organic and inorganic coatings. Prog Org Coat 151:106056
49. Velasco LF, Parra JB, Ania CO (2010) Role of activated carbon features on the photocatalytic degradation of phenol. Appl Surf Sci 256(17):5254–5258
50. Wei Z et al (2020) Hydrothermal synthesis and photo-Fenton degradation of magnetic $MnFe_2O_4$/RGO nanocomposites. J Mater Sci: Mater Electron 31(7):5176–5186. https://doi.org/10.1007/s10854-020-03077-4
51. Xu H-Y et al (2018) Heterogeneous Fenton-like discoloration of methyl orange using Fe_3O_4/MWCNTs as catalyst: kinetics and Fenton-like mechanism. Front Mater Sci 12(1):34–44. https://doi.org/10.1007/s11706-018-0412-5
52. Yang N et al (2019) Ternary composite of G-C_3N_4/$ZnFe_2O_4$/Fe_2O_3: hydrothermal synthesis and enhanced photocatalytic performance. ChemistrySelect 4(24):7308–7316. https://chemistry-europe.onlinelibrary.wiley.com/doi/abs/10.1002/slct.201901543
53. Yao Y, Zhao C, Zhao M, Wang X (2013) Electrocatalytic degradation of methylene blue on PbO_2-ZrO_2 nanocomposite electrodes prepared by pulse electrodeposition. J Hazard Mater 263:726–734. http://www.sciencedirect.com/science/article/pii/S0304389413007802
54. Yao Y et al (2016) Fe, Co, Ni nanocrystals encapsulated in nitrogen-doped carbon nanotubes as Fenton-like catalysts for organic pollutant removal. J Hazard Mater 314:129–139. http://www.sciencedirect.com/science/article/pii/S0304389416303193
55. Zhang X et al (2019) Enhancement of photo-Fenton-like degradation of orange II by MnO_2/NiO nanocomposite with the synergistic effect from bisulfite. J Alloys Compd 785:343–349. http://www.sciencedirect.com/science/article/pii/S0925838819302063
56. Zhao W et al (2020) PH-controlled $MnFe_2O_4$@ SnS2 nanocomposites for the visible-light photo-Fenton degradation. Mater Res Bull 124:110749. http://www.sciencedirect.com/science/article/pii/S0025540819330247
57. Zhou Y et al (2020) Enhanced degradation of acid Red 73 by using cellulose-based hydrogel coated Fe_3O_4 nanocomposite as a Fenton-like catalyst. Int J Biol Macromol 152:242–249. http://www.sciencedirect.com/science/article/pii/S0141813020305195

TiO_2-Based Composites for Water Decolorization

Indu M. Sasidharan Pillai and K. L. Priya

Abstract Photocatalysis, being a sustainable and environmentally friendly technology with potential for low-cost applications, is one of the most researched methods for dye decolourization. Among the various photocatalysts available, TiO_2 is a highly efficient and stable photocatalyst. Even though reports of scientific studies on the use of TiO_2 dates to 1930s, wider attention and numerous studies on its application as a photocatalyst for environmental remediation started after the study by Fujishima and Honda in 1972. The use of pure TiO_2 as photocatalyst is limited due to its high band gap energy and recombination rate. Also, the photo excitation of TiO_2 is in the ultraviolet (UV) range and consequently, only 5% of the total solar energy could be utilized for photocatalysis with pure TiO_2. Another limiting factor affecting the efficiency of degradation is mass transport limitation and agglomeration of the photocatalyst. This chapter reviews composites with metals, non-metals, semiconductors, carbon derivatives and other support structure, which are developed to tackle the limitations of photocatalysis with pure TiO_2. Metal-doped photocatalysts aid in reduced recombination and visible light photocatalysis. Non-metal doping leads to band gap modification and redshift. Heterojunctions with semiconductors as photosensitizer or electron sinks are possible. Carbon derivative-based composites and support structure-based composites also are studied extensively. Of late, co-modified composites are gaining importance. Overall, this book chapter intends to take the readers through the journey of development and application of TiO_2-based photocatalysts for dye degradation with emphasis on composites.

Keywords Photocatalysis · Dye decolourization · Composites · Advanced oxidation process · Recombination · Metal doping · Non-metal doping · Heterojunction · Intermediate energy level

I. M. Sasidharan Pillai (✉) · K. L. Priya
Department of Civil Engineering, TKM College of Engineering, Kollam 691005, Kerala, India
e-mail: indums@tkmce.ac.in

K. L. Priya
e-mail: priyaram@tkmce.ac.in

© The Author(s), under exclusive license to Springer Nature Singapore Pte Ltd. 2021

S. S. Muthu and A. Khadir (eds.), *Novel Materials for Dye-containing Wastewater Treatment*, Sustainable Textiles: Production, Processing, Manufacturing & Chemistry,
https://doi.org/10.1007/978-981-16-2892-4_5

1 Introduction

Freshwater resources are shrinking day by day, mainly due to overexploitation and pollution. According to the *World Water Development Report (2020)* globally about 4 billion people live under severe physical water scarcity and almost one-fourth of the world face economic water scarcity [126]. Coupled with the increased awareness about health-related problems due to water pollution, various governments were seen to take stricter legal action to throttle the pollution.

Among the various pollutants, dyes are a major issue worldwide. More than 10,000 different varieties of dyes are produced around the world [106]. Discharge of dye-containing wastewater to land and water bodies leads to many environmental and health issues. Aesthetic pollution, organic loading in the receiving water bodies, interference with penetration of light into waterbodies affecting the aquatic flora and fauna, toxicity and carcinogenicity of dyes, etc., are a few in the list [27, 96, 97, 107]. Adsorption, coagulation, chemical treatment, microbial remediation, etc., are applied for the treatment of this wastewater with different degree of success [29]. But the complete degradation and removal of dyes remain a challenge.

Advanced oxidation processes (AOPs) are now gaining importance as a major technology for wastewater management, especially for the treatment of recalcitrant organics and natural organic matter [113]. The method is based mostly on the use of hydroxyl radical (the second-best oxidizing agent next to fluorine) for the oxidation. The non-specific nature of hydroxyl radical and its ability to degrade a wide variety of contaminants helps in the degradation of complex wastewaters by AOPs. The main pathways for hydroxyl generation in AOPs are by chemical reaction (H_2O_2/Fe^{2+}, H_2O_2/O_3, etc.), photolytic techniques (Photocatalytic oxidation, photoelectrochemical oxidation, etc.) or electrolytic techniques (electrochemical oxidation, electro-Fenton, etc.). Among the different AOPs, photocatalysis is the most widely researched option. Many metal-based semiconductors including TiO_2, ZnO, SnO_2, Fe_2O_3, WO_3, VO_4, MoO_4, CeO_4, CdS, ZnS have proved efficient in the degradation of dyes [20]. Among the various photocatalysts available, TiO_2 is a highly efficient, low cost and stable photocatalyst [13, 18, 45, 93, 102, 105]. Hence, TiO_2 and its various modifications are studied elaborately as photocatalysts for dye degradation. This chapter discusses about the general mechanism of photocatalysis followed by a brief description on the genesis and of photocatalysis using TiO_2 and its various modifications. Mechanism of degradation with modified TiO_2 is examined. The modifications are classified under different heads and application of modified TiO_2 under each classification in dye degradation is reviewed.

2 Mechanism of Photocatalysis

In semiconductors, the energy difference between the bottom of the conduction band and top of the valance band is known as band gap. The electron bound in the valance band will move to the conduction band upon acquiring energy greater than or equal

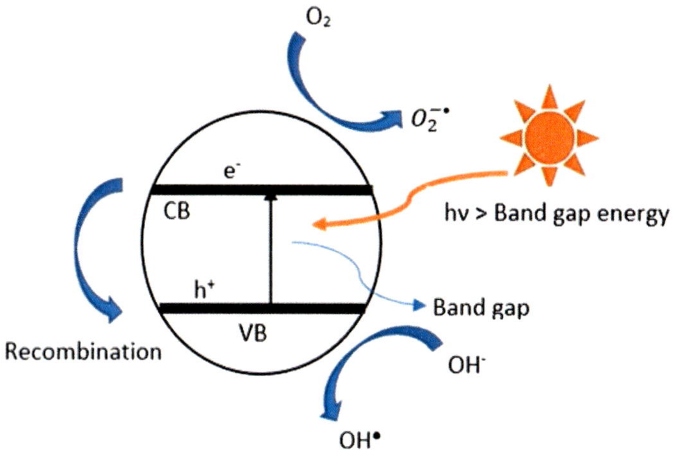

Fig. 1 Mechanism of photocatalytic oxidation with pure TiO_2

to the band gap. This electron is then free to move within the lattice structure of the crystal. In case of photocatalysis, photons of energy greater than the band gap excite the electrons in the valence band and the electrons move to the conduction band leaving a positively charged hole in the valence band (Eq. 1). This is commonly referred to as the generation of electron-hole pair [36].

The electron and hole migrate to the surface of the crystal, where it undergoes redox reactions with the adsorbed reactants. The hole can tap the electron from the hydroxyl ion in the water to produce hydroxyl radical (Eq. 2). This hydroxyl radical aids in the degradation of the contaminants in the bulk solution. But owing to the short lifetime of the hydroxyl radical, its action is limited to near the photocatalyst [37]. The electron can also generate superoxide radicals which aid in the degradation (Eq. 3). The detailed mechanism is depicted in Fig. 1.

$$TiO_2 + h\upsilon \rightarrow c_{cb}^{-} + h_{vb}^{+} \tag{1}$$

$$OH^{-} + h_{vb}^{+} \rightarrow H^{+} + {}^{\bullet}OH \tag{2}$$

$$O_2 + e_{cb}^{-} \rightarrow O_2^{-\bullet} \tag{3}$$

TiO_2 shows exceptionally good photocatalytic activity in the crystalline phase, whereas the activity in the amorphous phase is limited. TiO_2 exists in three crystal structures namely anatase, rutile, and brookite. The band gap energy for these three crystal structures is 3.2 eV, 3.0 eV and 3.54 eV for anatase, rutile and brookite structures, respectively [24, 31]. Of the three, anatase is more active photo catalytically owing to its better adsorption capacity, higher reduction potential and slower recombination [28, 48]. Degussa P25, which is one of the most widely studied commercial photocatalysts has anatase and rutile phase in the ratio 70:30 (approximately) and has better efficiency. This is due to reduced recombination owing to the formation of heterojunction between the two phases [31, 48].

Earlier studies on TiO_2 photocatalysis focused mostly on optimization of operating parameters. However, even though near 100% efficiency for dye removal was obtained at optimum conditions, the technology is not used in the field. The main challenges limiting the field application are listed below [13, 16, 19, 53, 54].

i. Photoexcitation of TiO_2 is in UV range owing to its high band gap energy. Since the UV component in solar light is less than 5%, this technology is not cost-efficient.
ii. The photonic efficiency of the system is reduced due to high electron-hole recombination (in the order of 10^{-9} s), whereas the degradation of contaminants takes 10^{-8} to 10^{-3} s, leading to low quantum yield.
iii. Agglomeration of TiO_2 nanoparticles causes a reduction in the surface area. Since photocatalysis is a surface phenomenon, this reduces efficiency.
iv. Transport of the contaminant from the bulk of the solution to the photocatalyst surface is required for the decomposition, i.e. the removal efficiency is mass transport controlled.
v. Separation of photocatalyst from the reaction medium after treatment.

In the last decade, researchers concentrated on overcoming these limitations of TiO_2 and as a result, many modifications, and composites of TiO_2 were prepared and tested. A brief history of the development of various photocatalysts is provided in the following section.

3 Genesis and Advancement of TiO_2 Photocatalysis

Reports of scientific studies on the use of TiO_2 dates to the 1930s. Wider attention and a large number of studies on its application as photocatalyst for environmental application started after the paper on water photolysis using TiO_2 by Fujishima and Honda in 1972 [94]. Environmental application of TiO_2 was first reported by Frank and Bard [94]. To tackle these issues and improve the properties of TiO_2 photocatalysts various modifications were carried out. The initial focus was to reduce the recombination and improve the photonic efficiency. One prominent modification in this aspect was the doping of the catalyst with various metals, especially noble metals.

Doping with transition metal elements, resulting in visible light-sensitive photocatalysis were studied extensively post the study by Anpo in the late 1990s [47]. However, e^- trapping by metal centres (beyond optimum doping), thermal instability and costly preparation process led researchers to explore non-metals for this purpose. Nitrogen is the most widely studied non-metal dopant, with a good visible light response [44]. The photocatalytic activity of nitrogen-doped TiO_2 was initially explored Sato in 1986 [47, 94]. Dye sensitized TiO_2 photocatalysis studied from the late 1990s also aided in visible light photocatalysis [47].

Immobilization on glass/quartz was attempted to separate the TiO_2 from the reactor. Various modifications like coating on magnetic iron oxide, different polymers, natural/biological membranes, etc., are attempted recently in this line [32, 46, 65, 98, 141].

Of late, most of the studies focus on modification with carbon-based materials. Another line of study is co-modification with two or more compounds to enhance specific properties of the photocatalyst. These are discussed in detail in the following sections.

4 Degradation Mechanism with Modified TiO_2

As evident from the previous section, many modifications are explored to improve the photocatalytic performance of TiO_2. These modifications improve the efficiency by different mechanisms. When materials with good electron affinity form composite with TiO_2, the excited electron from the conduction band is transferred to the element with high electron affinity. In other words, the material with good electron affinity acts as an electron sink, thereby preventing the recombination [1, 4, 49]. This helps in improving the photonic efficiency. The generalized mechanism is depicted in Fig. 2a.

Some elements in the composite can act as a photosensitizer, i.e. they can supply electron to the TiO_2 after photoexcitation (Fig. 2b). This can also result in visible light photocatalysis [78, 80]. Another mechanism of action is band gap modification (Fig. 2c). This is achieved by introducing intermediate energy levels between the valance band and conduction band by doping with nonmetals like nitrogen, carbon, etc., and metals like erbium [16, 17]. Reduced TiO_2 also has intermediate energy band [34, 76]. The reduction in band gap energy results in redshift and thus leads to visible light photocatalysis. Another mechanism cited for visible light photocatalysis is extrinsic adsorption owing to photoionization of defects induced by dopant elements [55, 79].

Composites of TiO_2 with metals and semiconductors result in the formation of heterojunction between the semiconductors (Fig. 2d, e). In a heterojunction free electron movement from the conduction band of the semiconductor at a higher energy level to the conduction band of semiconductor at lower energy level takes place. Depending on the type of semiconductor used in the composite, this can aid in visible light photocatalysis and/or reduction in recombination rate [57, 75, 88]. Transfer of holes from higher energy level valance band to lower energy level valance band is also possible [88].

Many carbon-based substrates act as electron sink and transport the electron away from the photocatalyst resulting in reduced recombination. They are also good adsorbents and hence the mechanism of action includes adsorption and photocatalytic degradation. In this type of photocatalysis mass transfer limitation, mentioned earlier is minimized (Fig. 2f).

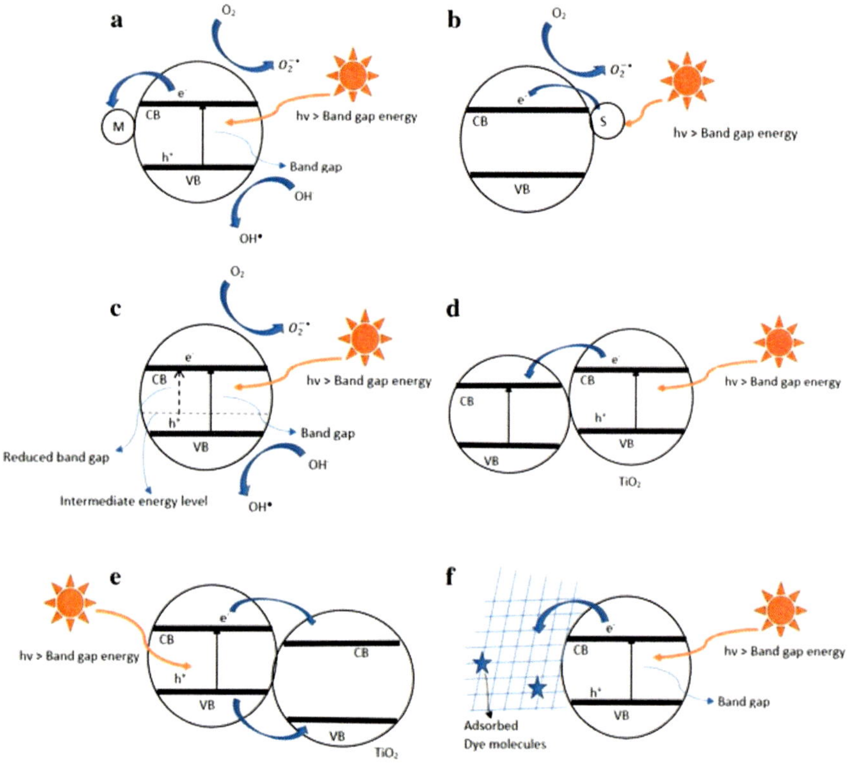

Fig. 2 Mechanism of action of TiO_2 composites **a** electron sink **b** photosensitizer **c** band gap modification **d** heterojunction with electron sink **e** heterojunction with photosensitizer **f** transport the electron and adsorption of contaminants

The subsequent sections of this chapter focus on the various TiO_2 composites applied for dye removal. A wide variety of complexes are developed for dye removal and hence it is important to classify them before moving on with the discussions.

5 Classification of TiO_2 Composites

A direct classification of the TiO_2 composite is not possible due to its varied nature. Dahl et al. [28] classified the TiO_2 composites based on their position in the periodic table. A pictorial representation of the classification proposed by them is shown in Fig. 3a. The classification by Han et al. [44] is also similar to Dahl et al. [28], but along with the classification based on the periodic table, functional classification was also done. The details are provided in Fig. 3b. A more concise classification was followed by Nasr et al. [84] in their review (Fig. 3c).

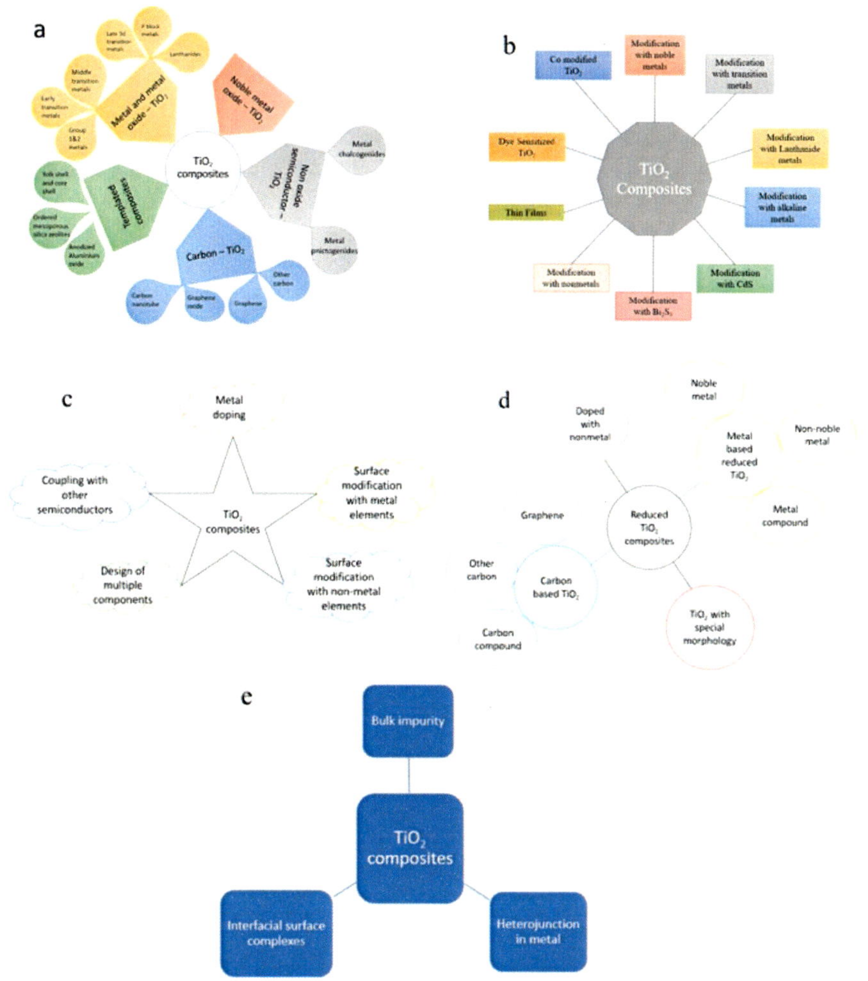

Fig. 3 Classification of TiO_2 composites based on literature **a** Dahl et al. [28] **b** Han et al. [44] **c** Nasr et al. [84] **d** Fang et al. [34] **e** Higashimoto [47]

Fang et al. [34] prepared a generalized classification for visible light active reduced TiO_2 as non-metal doped TiO_2, metal-based reduced TiO_2, carbon-based TiO_{2-x} and TiO_{2-x} with special morphology and is represented in Fig. 3d. Higashimoto [47] also represented visible light active photocatalysts under generalized heads as given in Fig. 3e.

Based on the above classifications and review of other articles, TiO_2 composites are classified under the following in the present study.

1. Metal/metal oxide-based composites
2. Non-metal-based composites
3. Carbon derivatives-based composites

4. Semiconductor-based composites
5. Support structure-based composites
6. Co-modified composites

Both metal and non-metal-based composites have improved properties due to bulk modification of the TiO_2 crystal, whereas the rest achieves improvement in the property by surface modification [63].

5.1 Metal/Metal Oxide-Based Composites

Studies on decolourization of dye using alkali metals and alkaline earth metals are limited. TiO_2 modified with Na, Li and K were studied by Bessekhouad et al. [15]. Both Na and K incorporation led to the deactivation of the photocatalyst in the degradation of malachite green oxalate and 5% Li impregnated TiO_2 performed better than undoped TiO_2 in the degradation of malachite green oxalate. Crystallinity was reported to increase with an increase in Li concentration, whereas K doping had a detrimental effect on crystallinity. On the contrary, Chen et al. [22] reported an improvement in photocatalytic efficiency with K^+ modified composite at optimum preparation conditions. In general, alkali metals are not an attractive dopant of TiO_2.

Transition metals are used widely for the enhancement of the photocatalytic properties of TiO_2. Most of these metals act as e^- sinks, thus preventing recombination. In W(IV) doped TiO_2, the improvement in the efficiency of decolourization of crystal violet under UV light was attributed to the e^- sink action of W(IV) [26]. Lee et al. [65] also reported that W doped TiO_2 performed better in the degradation of methylene blue under UV light. They compared the performance of W doped, Al-doped and (W + Al) doped TiO_2 prepared by dip-coating method and observed that the performance of Al-doped TiO_2 was poorer than pure TiO_2.

Silver doped TiO_2 prepared by sol–gel method showed 99.5% efficiency in decolourization of methyl orange under UV irradiation. The authors suggested that the improvement in efficiency is owing to the e^- sink action of Ag^+ [122]. Reduction in recombination by Ag^+ doping of TiO_2 was observed with Ag-TiO_2 thin films deposited on Silica substrate [30]. Pd also improves the photocatalytic decolourization efficiency by e^- sink mechanism [87].

Some of the transition metals also move to the crystal lattice and result in band gap modification and improves the visible light excitement. Fe^{3+} with a similar ionic radius as Ti^{4+} is one of the easiest metals to incorporate to TiO_2 crystal lattice. They are reported to create new energy level and act as an e^- sink [7, 32]. A higher concentration of Fe^{3+} is reported to reduce the crystal size. However, due to particle agglomeration, the surface area was lesser compared to Degussa P25 [7]. Similar observation on reduction on particle size at higher iron doping was provided by [5]. Momeni and Ghayeb [79] developed visible light active Co-TiO_2 nanotube by photo-assisted deposition of cobalt. The resulting composite had a band gap of 2.42 eV and exhibited around 70% decolourization efficiency for methylene blue under visible light irradiation. Ni doping is also reported to be quite efficient in improving the visible light response of TiO_2, with more than 80% methylene blue removal efficiency under UV irradiation with Ni-doped TiO_2 [55]. In both the studies, the removal efficiency by pure TiO_2 was reported to be less than 10% under similar degradation conditions. Doping of Ce ions to TiO_2 lattice was reported by Yan et al. [135]. A red shift and visible light activity were reported with Ce doping. Chen et al. [23] also reported visible light activity of Ce-doped TiO_2. However, they had reported that Ce is attached to the surface of TiO_2 and cannot move to the bulk of the crystal. The contradiction could be due to the difference in preparation conditions. Nearly 100% removal efficiency for methyl orange was obtained with Cu-doped TiO_2 having the optimal doping of 1 mol%.

Heterojunctions also aid in improving photocatalytic efficiency. Bi-TiO_2 nanotubes were reported to form heterojunction, where Bi acted as an e^- sink. Band gap energy reduction from 3.18 eV (pristine TiO_2) to 2.82 eV was also reported leading to 3.6 fold improvement in methylene blue degradation under visible light [4]. Heterojunction of TiO_2 film with silver ion was reported by Demirci et al. [30].

The effectiveness of the modification depends on several factors including the relative proportion of the metal/metal oxide in the composite, calcination, technique used for the modification, etc. If the dosage of the metal is beyond an optimal limit, they can act as centres of recombination, thus affecting the efficiency detrimentally [32, 44, 138]. The optimal concentration for Ag^+ was reported as 0.25 mol% [122], whereas for another noble metal, Pt, the optimum doping was reported to be 0.5 atom% [56] and for Pd, it was 0.5 wt% [87]. Fe^{3+} was also reported to function efficiently at low molecular concentration, the efficiency reducing with an increase in concentration [32]. In the case of La, 1 mol% was reported as the optimum doping [69]. Cu^+ also was reported to perform best at 1% mole fraction [138]. Study on various rare-earth-doped TiO_2 (La^{3+}, Ce^{3+}, Ce^{4+}, Nd^{3+}, Sm^{3+}, Pr^{3+}, Eu^{3+}, Gd^{3+} and Dy^{3+} doped TiO_2) showed a red shift of the doped composites with optimum doping in the range of 0.5–1 weight % [120].

Chen et al. [22] reported that the efficiency of the photocatalyst depends on the preparation conditions. An improvement in photocatalytic efficiency was observed for TiO_2 with 4.6% mole fractions of K^+ and calcined at 700 °C, whereas the efficiency was low for composites calcined at a lower temperature. In the modification of TiO_2 with alkali metals impregnation method was better than the sol–gel preparation method [15]. An increase in the concentration of Tungsten (IV) retards conversion from anatase phase to rutile phase and hinders crystal growth. The optimum concentration of W(IV) for improving the photocatalytic efficiency is reported as 3% [26]. A similar mechanism was also reported for Ce doped TiO_2. Ce inhibits the conversion of anatase to rutile phase and restrain the growth of TiO_2 particle size [135]. A reverse trend was observed for V doped TiO_2, where the anatase to rutile conversion happened at 300 °C [143].

As evident from the discussion, it is extremely critical that the doping is performed at an optimal concentration. Beyond the optimal concentrations the metal act as centres of recombination, thereby reducing the efficiency. Thermal instability of the modified photocatalysts is also an issue. Another pertinent issue with metal-doped photocatalyst is the possibility of metal leaching. Details of different metal-doped composites are provided in Table 1.

5.2 Non-metal-Based Composites

Non-metal doping in general has minimal effect as recombination centres, which was one of the major limitations of metal-doped TiO_2 [21]. Nitrogen is one of the widely studied non-metal dopants. Nitrogen doping can function in two ways i.e. while substitution of lattice oxygen by N results in band gap narrowing, interstitial nitrogen results in the formation of a mid band gap. In both cases, the band gap energy is reduced and the reduction is more in case of the latter [21, 47]. N-TiO_2 prepared by a low-temperature hydrothermal method had the presence of both interstitial and substitutional nitrogen [133]. Le et al. [64] also reported the incorporation of nitrogen by substitutional and interstitial means where substitutional doping resulted in a band gap of 3.06 eV and interstitial doping lead to a band gap of 2.46 eV. In N-TiO_2 prepared by chemical precipitation followed by calcination, the substitution of lattice oxygen by N was observed and the reduction in band gap was estimated as 0.14 eV. N-TiO_2 thus developed performed better than Degussa P25 in the visible region, while in the UV range P25 performed better than N-TiO_2.

Carbon is another major non-metal dopant efficient in imparting visible light photocatalytic activity of TiO_2. According to Chen et al. [21], C atom has difficulty in diffusing to TiO_2 bulk of TiO_2 crystal and rather exists on the surface and suppresses crystal growth and the carbon species on the surface of TiO_2 lead to agglomeration of the photocatalyst. On the contrary, Xiao et al. [130] indicated that carbon substitutes some of the Ti atom in the crystal lattice resulting in the formation of reduced Ti and consequent red shift. Substitution of oxygen in the crystal lattice by carbon was also observed [112, 128]. Mechanism of photodegradation with C doping was suggested to be photosensitization by the carbon species on the surface of TiO_2 by Chen et al. [21]. Band gap modification by reducing Ti was suggested as the reason for enhanced visible-light activity in C-TiO_2 prepared by sol–gel auto-combustion technique [130]. Carbon also improved the adsorption of dyes and thus reduced the mass transfer limitation [104, 112].

Sulphur is also reported to enhance visible light activity. During doping of TiO_2 with cationic sulphur, the dopant substituted Ti from the crystal lattice causing a red shift owing to band gap modification [14, 83, 90]. Band gap modification by substitution of oxygen in the lattice with anionic sulphur was also studied [125]. Doping process did not affect the crystal phase of TiO_2 [14]. F–S-co-doped TiO_2 showed better performance than F doped TiO_2 and S doped TiO_2. The co-doped composite had excellent structural stability [137]. The ionic radius of fluorine and oxygen are remarkably close and hence oxygen in the TiO_2 lattice can be replaced by Fluorine with ease. Inhibition of anatase to rutile crystal transformation was also reported [131]. Fluoride doping also has the advantage of suppression of brookite crystal formation [140]. Fluorine doping does not produce mid-gap energy levels but

Table 1 Preparation procedure and performance of metal doped composites

Catalyst	Preparation procedure	Light source	Dye/Efficiency	Improvement/Modification	Recyclability	References
Bi-TiO_2	Hydrothermal synthesis	Visible light	Methylene blue/3.6 times improvement	Visible light photocatalysis Reduced recombination	Reduction in efficiency from 64 to 57% after 5 cycles	Ali and Kim [4]
Ni-TiO_2	Hydrothermal method followed by calcination	Visible light	Methylene blue/Approximately 80% efficiency	Visible light photocatalysis	Negligible reduction in efficiency after 4 cycles	Kianfar et al. [55]
Ag-TiO_2	Microwave-assisted method	UV light	Methyl orange/99.5%	Reduce recombination	20% reduction in efficiency after 5 cycles	Suwarnkar et al. [122]
Ce-Si/TiO_2	Sol–gel + calcination	Visible light	Rhodamine B	Visible light photocatalysis Thermal stability	–	Chen et al. [23]
Ce-TiO_2	Sol–gel	Visible light	Methylene blue/90.03%	Visible light photocatalysis Thermal stability	–	Yan et al. [135]
Co-TiO_2	One-step electrochemical anodizing + photo-assisted deposition of Co	Visible light	Methylene blue/70%	Visible light photocatalysis	No reduction in efficiency after 4 cycles	Momeni and Ghayeb [79]
V-TiO_2	Sol–gel method	Visible light	Methyl orange/approximately 30%	Visible light photocatalysis	–	Zhou et al. [147]

(continued)

Table 1 (continued)

Catalyst	Preparation procedure	Light source	Dye/Efficiency	Improvement/Modification	Recyclability	References
Fe^{3+}-TiO_2	Hydrothermal process	UV light and UV-Vis light	Malachite green/80% (approximate) under UV Irradiation and 70% (approximate) under visible light	Visible light photocatalysis Reduced recombination	–	Asiltürk et al. [7]
Cu-TiO_2	Sol–gel method	UV-Vis light	Methyl orange/~100%	Enhanced photonic efficiency Reduced recombination	–	Yang et al. [138]
Pd-doped TiO_2	Sol–gel method	UV light	Methylene blue/100% Methyl orange/100%	Enhance light harvesting Reduce recombination	More than 95% removal efficiency after 10 cycles	Nguyen et al. [87]
Ag-doped TiO_2 film	Sol–gel spin coating technique on Si substrates	UV-Visible light	Methylene Blue/approximately 50%	Reduce recombination	–	Demirci et al. [30]
La^{3+}, Ce^{3+}, Ce^{4+}, Nd^{3+}, Sm^{3+}, Pr^{3+}, Eu^{3+}, Gd^{3+} and Dy^{3+} doped TiO_2	Homogeneous hydrolysis using urea as the precipitation agent	UV and visible light	Orange (II)	Visible light activity Reduced recombination	–	Štengl et al. [120]

leads to the formation of surface oxygen vacancies which is important for visible light absorption [137]. Fluorine doping affects the strength and mode of dye adsorption and due to the surface charge modification associated with fluorine doping retardation in degradation efficiency of anionic dyes are noted [44].

Other anions doped with TiO_2 include boron, phosphorus, iodine, etc. [67, 117, 136]. In all the cases, visible light photocatalysis is achieved. In general, non-metal doping is a band gap engineering technique where redshift in light absorption is achieved mainly through the introduction of sub-levels or creation of oxygen defects. Despite the visible light activity, other limitations of TiO_2 photocatalysis remain in non-metal modified catalysts and need to be addressed. Table 2 summarizes different nonmetal doped composites.

Calcination temperature plays a crucial role in determining the efficiency of nonmetal doped TiO_2. Le et al. [64] reported that the dopant (N) concentration reduced with an increase in calcination temperature. Agglomeration of particles with the increase in temperature was also reported. The optimum calcination temperature was 400 °C [110]. N-TiO_2 prepared by sol–gel method and calcined at 400 °C proved to be an excellent photocatalyst in the degradation of azo dyes. The study stresses that the photocatalyst remains in the amorphous state before calcination and is less efficient [61, 70]. The effect of calcination temperature on carbon doping was studied in detail and 600 °C was found to be optimal [130]. A loss in carbon content with an increase in the calcination temperature was reported by Matos et al. [77]. Temperature is reported to have a crucial role while doping TiO_2 with sulphur and the optimum temperature obtained was 500 °C [14, 83, 90]. Carbon doping is found to inhibit the formation of Brookite TiO_2 [77, 104]. A reduction in crystal size with doping was also reported [21, 70, 110]. The phase transformation from anatase to brookite was also inhibited by N-doping [25]. Unlike metal doping optimum concentration of the dopant was not considered in most of the studies, however Le et al. [64] reported the optimum N concentration obtained in their study as 9.47 atomic mass%.

5.3 Carbon Derivatives-Based Composites

In this section, a discussion on various composites based on different forms of carbon like activated carbon, carbon nanotube, fullerene (C60), graphene, graphene oxide (GO), reduced graphene oxide (rGO), xerogel, etc., which is extensively studied recently, is included. Optical and photocatalytic activities can be significantly improved in the case of nanocarbon derivative-based composites [17]. Table 3 indicates the performance of TiO_2/carbon derivates-based composites. Unlike carbon doping, these carbon derivatives modify the properties of TiO_2 by different mechanisms and in general, are more advantageous. They act as conductive linkers, e^- sink, photosensitizers, and support for catalysts [9, 13]. The simplest of these modifiers is activated carbon-based composites where the activated carbon acts as a support for TiO_2 and adsorb pollutant from the bulk of the solution and concentrates it near the TiO_2 surface [6]. The photodegradation by commercial activated carbon/TiO_2

Table 2 Preparation procedure and performance of non-metal doped composites

Catalyst	Preparation procedure	Light source	Dye/Efficiency	Improvement/Modification	Recyclability	References
N-TiO_2	Chemical precipitation followed by calcination	UV light Visible light	Methylene blue/approximately 10% decolourization efficiency in 400–450 nm wavelength region	Red shift and visible light absorption	–	Sathish et al. [110]
N-TiO_2	Sol–gel method	UV light Solar light	Acid orange 7, Procion red MX-5B and Reactive black 5/100% efficiency under UV light (2 h) and sunlight (3 h)	Visible light photocatalysis	–	Liu et al. [70]
N-TiO_2	Sol–gel method	Visible light	Methylene blue/approximately 60%	Visible light photocatalysis	–	Chen et al. [21]
C-N-TiO_2	Sol–gel method	Visible light	Methylene blue/approximately 65%	Visible light photocatalysis	–	
C-TiO_2	Sol–gel method	Visible light	Methylene blue/approximately 90%	Visible light photocatalysis	–	
N-TiO_2	Hydrolysis and calcination	Visible light Sunlight	Orange G/approximately 100% degradation under both light	Visible light photocatalysis	–	Sun et al. [121]

(continued)

Table 2 (continued)

Catalyst	Preparation procedure	Light source	Dye/Efficiency	Improvement/Modification	Recyclability	References
N-doped TiO_2 nanotube arrays	Annealing process at 450°C	Visible light	Methylene blue/approximately 50%	Visible light photocatalysis	–	Le et al. [64]
C-TiO_2	Sol–gel auto-combustion technique	Sunlight	Methylene blue/100% in 1 h	Visible light photocatalysis	–	Xiao et al. [130]
Mesoporous C-TiO_2	Hydrothermal method at low temperature	Visible light	Rhodamine B/approximately 60%	Visible light photocatalysis Mass transfer limitation is reduced by adsorption		Ren et al. [104]
C-TiO_2	Solvothermal synthesis and by calcination	UV, UV-vis, and Visible light	Methylene blue/100% under UV light for 210 min, 100% under UV-Vis light for 300 min and 90% under visible light for 300 min	Visible light photocatalysis	–	Matos et al. [77]
C-doped TiO_2 single-crystal nanorods	Hydrothermal synthesis	Visible light	Methylene blue/100% in 1 h Rhodamine B/100% in 20 min	Visible light photocatalysis Mass transfer limitation addressed	A Negligible reduction in efficiency after 3 cycles	Shao et al. [112]
Black carbon doped TiO_2 films	Unbalanced magnetron sputtering technique	UV(A) Visible light	Methylene blue/good removal with UV(A) and visible light irradiation	Visible light photocatalysis	–	Varnagiris et al. [128]

(continued)

Table 2 (continued)

Catalyst	Preparation procedure	Light source	Dye/Efficiency	Improvement/Modification	Recyclability	References
S cation-TiO_2	Sol–gel process	Simulated solar light	Methylene blue/decolourization efficiency better than pure TiO_2 at wavelengths above 440 nm	Visible light photocatalysis	–	Ohno et al. [90]
S-doped TiO_2	Thermal hydrolysis method	Visible light	Methylene blue/approximately 70%	Visible light photocatalysis	–	Nam et al. [83]
S-TiO_2 film	Metal-organic chemical vapour deposition	Visible light Sunlight	Methyl orange/approximately 40% in 5 h	Visible light photocatalysis Reduced recombination	–	Bento et al. [14]
F-doped TiO_2 film	Hydrolysis of titanium-n-butoxide in abundant NH_4F-H_2O acidic solution	Artificial solar light	X-3B/approximately 80%	Visible light photocatalysis	No reduction in efficiency after 6 cycles	Xu et al. [131]
F–S co-doped TiO_2	Low-temperature solvothermal method	Visible light	Methylene Blue/more than 95%	Visible light photocatalysis	–	Yang et al. [137]
B-TiO_2	One-step anodization	UV-Visible light	Methylene blue/more than 90%	Visible light photocatalysis	–	Siuzdak et al. [117]
I-TiO_2	Sol–gel	Sunlight	Rhodamine B/approximately 95%	Visible light photocatalysis	A Negligible reduction in efficiency after 3 cycles	

Table 3 Preparation procedure and performance of TiO_2/carbon derivates-based composites

Catalyst	Preparation procedure	Light source	Dye/Efficiency	Improvement/Modification	Recyclability	References
Activated Carbon-TiO_2	Metal-organic chemical vapour deposition	UV light	Tartrazine/96%	Mass transport limitation addressed Separation from the reaction medium Aggregation issue of TiO_2 addressed	30% reduction in efficiency after 4 cycles	Andriantsiferana et al. [6]
Activated Carbon-TiO_2	Sol–gel technique	UV light	Acid red 131/approximately 90%	Mass transport limitation addressed Separation from the reaction medium	–	Singh et al. [114]
Activated Carbon-TiO_2	Sol–gel technique	UV light	Sunset yellow/more than 90%	Mass transport limitation addressed Separation from the reaction medium	6% reduction in efficiency after 4 cycles	Rajamanickam and Shanthi [99]
GO-TiO_2	Hydrothermal method followed by calcination treatment	UV-visible light	Rhodamine B/100% Acid green 25/96%	Reduce recombination Mass transport limitation addressed	After 4 cycles around 9% reduction in efficiency	Adly et al. [1]

(continued)

Table 3 (continued)

Catalyst	Preparation procedure	Light source	Dye/Efficiency	Improvement/Modification	Recyclability	References
GO-TiO_2	Hydrothermal method	Sunlight Artificial solar light	Congo red/approximately 90% Methylene blue/80-90%	Visible light photocatalysis Reduced recombination Mass transport limitation addressed	10% reduction in efficiency for congo red and 3% reduction in efficiency for methylene blue after 5 cycles	Alamelu et al. [3]
GO-TiO_2	Solvothermal method	UV light	Methyl orange/84% in 240 min Methylene blue/100% in 25 min	Reduced recombination	–	Atchudan et al. [8]
rGO/TiO_2	Hydrothermal process	UV light and visible light	Methylene blue/100% under UV and more than 90% under visible light	Visible light photocatalysis Reduced recombination Mass transport limitation addressed	–	Nguyen and Juang [86]
Graphene/TiO_2	Microwave-assisted synthesis	UV light	Methylene blue/approximately 80% Rose Bengal/approximately 70%	Reduced recombination Mass transport limitation addressed	–	Gayathri et al. [39]
Graphene/TiO_2	Sol–gel method	Artificial solar light Visible light	Methylene blue/complete removal	Visible light photocatalysis Reduced recombination Mass transport limitation addressed	–	Liu et al. [71]

(continued)

Table 3 (continued)

Catalyst	Preparation procedure	Light source	Dye/Efficiency	Improvement/Modification	Recyclability	References
Carbon nanotube/TiO_2 nanotube	Incorporation of Carbon nanotube in HNO_3 washing process	UV light	Methyl orange/94%	Reduced recombination Mass transport limitation addressed	Efficiency reduced by 7.6% in three cycles	Jiang et al. [50]
MWCNT/TiO_2	Sol–gel method	UV light	Methyl Orange/less efficient than P25 and MWCNT	Reduced recombination Mass transport limitation addressed	–	Da Dalt et al. [27]
MWCNT/TiO_2	Solvothermal method	UV light	Methyl orange/more than 90%	Reduced recombination Mass transport limitation addressed	–	Saleh and Gupta [108]
Carbon xerogel/TiO_2	Sol–gel method	Visible light	Orange G/90% at 400 min	Visible light photocatalysis Reduced recombination Mass transport limitation addressed	–	Bailón-García et al. [13]

composite is reported to be two times faster than that by pure TiO_2 owing to the highly porous nature of the commercial activated carbon [82]. More than 90% dye removal efficiency for activated carbon/TiO_2 composite was reported under UV irradiation [99].

Single-walled carbon nanotube (SWCNT) reduces recombination by e^- sink and transport action and hence the performance of TiO_2-SWCNT ($k = 2.4 \times 10^{-3}$) was reported as better than TiO_2 alone ($k = 9.7 \times 10^{-4}$) [19]. Da Dalt et al. [27] prepared multiwalled carbon nanotube (MWCNT)/TiO_2 composites by sol–gel method, but the performance of the prepared composite was poorer than pure TiO_2 and MWCNT. Oh et al. [89] performed a comparative assessment of the efficiency of two carbon-based photocatalysts in the degradation of methylene blue and concluded that the performance of MWCNT/TiO_2 (1.19 times more efficient than pristine TiO_2) was better than that of C_{60}/TiO_2 (1.03 times more efficient than pristine TiO_2). In another comparative assessment study, Degussa P25/graphene composite performed better than Degussa P25/carbon nano tube composite under both UV light and visible light. Performance of both the composites was much better than pure TiO_2 [141].

As in the case of carbon-doped TiO_2, a detailed analysis of XRD spectra of carbon nanotube/TiO_2 indicated that TiO_2 crystals are not well developed (amorphous TiO_2). The phase conversion was achieved by thermal treatment [27]. Anatase crystal structure formed at a temperature range of 350–450 °C and for a temperature of 550 °C, rutile phase was prominent. Surface area also decreased with the increase in temperature [50]. Another factor affecting the degradation efficiency was the mass ratio of carbon nanotube. At lower mass ratios (0.05 and 0.1) improvement in the efficiency of the composite was observed. However, at a mass ratio of 0.15 and above, the efficiency of the composite was less than that of TiO_2 nanotube, which was attributed to photon scattering by surplus carbon [50]. Other forms of carbon also show a similar trend. Reduced graphene oxide/TiO_2 composite at 3% rGO concentration had 4.1 fold higher photocurrent in contrast to TiO_2 nanotube [86]. The optimal mass ratio of graphene oxide in GO/TiO_2 was reported as 10% [85]. In the study conducted by Song et al. [118], a higher mass ratio of 15% was reported to be optimal. Both crystal phase and size are affected by carbon concentration [13, 41].

Graphene sheets form a heterojunction with TiO_2 and act as acceptors and transporters of photogenerated electrons and suppress electron-hole recombination [17, 81]. This property resulted in 12 times better photocatalytic activity of graphene oxide-TiO_2 nanocomposites than nano TiO_2. However, a significant difference in time required was observed for the degradation of methyl orange (240 min) and methylene blue (25 min), indicating the selective nature of the photocatalyst [8]. Apart from acting as an e^- sink, carbon composites also act as a sensitizer [71, 78]. The sensitizer effect of reduced graphene oxide was observed under visible light, whereas under UV light it acts as an electron acceptor [93]. Functionalized carbon nanotube and carbon xerogel can also act as a sensitizer [13, 109].

The preparation procedure has a major role in determining the photocatalytic efficiency of the composite. Sampaio et al. [109] reported that TiO_2/carbon nanotube composite prepared through sol–gel method had improved efficiency in comparison with pure TiO_2, whereas for TiO_2/carbon nanotube composite prepared through hydration-dehydration technique the efficiency was lesser than that of pure TiO_2. The properties of the composites also varied with the preparation procedure.

A major limitation of the carbon-based composites is the masking of light by carbon, i.e. weakening of the light intensity arriving at the surface of the catalysts due to the presence of nanocarbon and lack of reproducibility of the photocatalyst due to variations in preparation and treatment [1, 9, 92]. Table 3 indicates good recyclability for the carbon-derivatives-based TiO_2 photocatalysts.

5.4 Semiconductor-Based Composites

In semiconductor-based composites, two semiconductors interact over a heterojunction resulting in reduced recombination, visible light photocatalysis, etc., based on the property of the semiconductors. Different semiconductor composites were attempted to improve the property of the TiO_2. Fang et al. [35] developed Cu_2O/TiO_2 composite with visible light activity. They reported that the calcination temperature had an important role in the crystallization with the temperature between 400–600 °C yielding high purity anatase TiO_2. The band gap of Cu_2O is 2.2 eV and hence photoexcitation occurs in the visible wavelength. The excited electron is transferred to the conduction band of TiO_2, aiding degradation. Thus. the composite has reduced recombination and improved light response [35]. CuO/TiO_2 composite was also successfully attempted. Reduction in band gap energy with an increase in CuO concentration was reported [57]. Composite of TiO_2 with WO_3 was also reported to perform well under visible light and sunlight [72]. The composite has a reduced band gap (2.78 eV) and improved efficiency [79]. Composite with SeO_2 also results in visible light photocatalysis [100]. ZnO is another semiconductor used to form composite with TiO_2. The band gap of ZnO is larger than that of TiO_2 and the composite has reduced recombination [127]. SnO_2 is reported to stabilize the anatase phase of TiO_2. It also reduces recombination and improves solar light photocatalytic activity by reducing the band gap [59, 101, 111]. The optimum concentration of SnO_2 in the composite was reported as less than 10% [59].

$TiO_2/ZnFe_2O_4$ was found to be 4–5 times more efficient than pure TiO_2 in visible light photocatalysis of methyl orange. The magnetic property of the composite also aids in easy separation from the reactor [132, 134]. TiO_2/magnetic Fe_3O_4 composite prepared by acid sol method had the dual advantage of easy separation of photocatalyst from the reactor and improved decolourization efficiency owing to e^- sink action of Fe^{3+}. The composite achieved complete decolourization of reactive brilliant red X-3B under UV irradiation [24]. The study highlights the importance of the preparation method on the efficiency of magnetic nanoparticles.

Other than metal oxides, metal sulphide semiconductors were also effectively used for the preparation of composite with TiO_2. Composite of MoS_2 with TiO_2 was 2.5 times more efficient than pure TiO_2. The semiconductor was reported to form a heterojunction with TiO_2 and reduce recombination with MoS_2 acting as an electron sink [49]. In another study MoS_2 nanosheet-mesoporous TiO_2 composite prepared by hydrothermal approach was shown to perform well under sunlight with greater than 90% colour removal efficiency while mesoporous TiO_2 had around 65% efficiency. The optimum proportion was 0.3 weight % of mesoporous TiO_2 on MoS_2 nanosheet [53]. Visible light photocatalysis of drimaren red, rhodamine B and methylene blue with CdS/TiO_2 was reported [11, 40, 66, 124]. The optimal proportion of CdS to TiO_2 was reported to be 12% by Li et al. [66], whereas TiO_2/CdS ratio of 4:1 was reported to be optimal by Bai et al. [12].

Composite with amorphous carbon nitride, a semiconductor in the visible region with band gap of 2.7 eV is reported to have high efficiency in comparison to commercially available TiO_2 [9]. Simonkolleite, a wideband semiconductor with band structure similar to ZnO, was used to prepare composite with TiO_2 and the prepared composite had visible light activity [10, 91]. A considerable reduction in the surface area upon the formation of the composite was reported [10]. The details of various semiconductor-based composites are shown in Table 4.

5.5 *Support Structure-Based Composites*

Support structure-based TiO_2 composites address three different limitations, individually or in conjunction, i.e. mass transport limitation, agglomeration of photocatalyst and recovery of photocatalyst after the reaction. Some support structures also interact with the photocatalyst resulting in visible light photocatalysis and/or reduced recombination. As described in Sect. 5.3, carbon derivatives like activated carbon, carbon nanotube, etc., act as efficient support structures for TiO_2 photocatalysis with good adsorption capacity and dispersion of TiO_2 [108, 114].

Researchers addressed the issue of difficulty in the separation of TiO_2 from the reaction mixture majorly in two ways. One method is the preparation of composites with a magnetic nanoparticle. The second method is the support structure-based composites. Earlier glass and quartz were tried as support structures [32, 65, 119]. TiO_2 embedded on poly(methyl methacrylate) was efficiently used to retrieve nano photocatalyst from the slurry reactor [19]. Floating photocatalyst was developed by using low-density polyethylene as the substrate. The composite had a strong bonding and leaching was not reported, however, agglomeration of the photocatalyst at high TiO_2 concentration was reported as an issue [74]. Poly sulfone is another polymeric membrane attempted for the immobilization of TiO_2 [60]. Composite with a conductive polymer (polypyrole) had good decolourization efficiency under UV light and sunlight. The conductive polymer acted as a photosensitizer in the composite, but the reusability of the composite is not good [58]. Poly [acrylamide-co-(acrylic acid)] also served as good support at low TiO_2 concentration. But agglomeration at high TiO_2 concentration and poor recyclability due to change in morphology of the hydrogel during photodegradation are the major limitations [51]. Methyl cellulose, porous poly dimethyl siloxane, bentonite, poly aniline, etc., are other supports studied [43, 46, 62, 103]. An added advantage of these composites is its increased porosity and improved adsorption capacity which helps in addressing the mass transport limitation

Table 4 Semiconductor-based TiO_2 composites in dye degradation

Catalyst	Preparation procedure	Light source	Dye/efficiency	Improvement/Modification	Recyclability	References
ZnO/TiO_2	Sol–gel	Visible light	Methylene blue/more than 80%	Reduced recombination Extended light absorption to visible range	–	Upadhyay et al. [127]
TiO_2/SnO_2 Nanotube	Hydrothermal preparation followed by chemical precipitation	UV light and Sunlight	Textile wastewater/more than 80%	Reduced recombination Extended light absorption to visible range	1.31% reduction in efficiency after 5 cycles	Rajkumar et al. [101]
SeO_2/TiO_2	Sol–gel	Sunlight	Sunset yellow/more than 90%	Reduced recombination Extended light absorption to visible range	2% reduction in efficiency after 4 cycles	Rajamanickam et al. [100]
TiO_2/Cu_2O	Sol–gel	Visible light	Reactive brilliant red X-3B/more than 90%	Extended light absorption to visible range Reduced recombination	–	Fang et al. [35]
TiO_2/CuO	Sol–gel	UV light	Methyl orange/80% approximately	Extended light absorption to visible range	–	Koohestani and Sadrnezhaad [57]
TiO_2/Fe_3O_4	Acid–sol method homogeneous precipitation method	UV light	Reactive brilliant red X-3B/more than 95% for both preparation methods	Easy separation of photocatalyst	–	Chen et al. [24]
$ZnFe_2O_4/TiO_2$	Microemulsion method	Visible light	Methyl orange/90% approximately Rhodamine B/90% approximately	Reduced recombination Extended light absorption to visible range Easy separation of photocatalyst	Recyclable	Xu et al. [134]

(continued)

Table 4 (continued)

Catalyst	Preparation procedure	Light source	Dye/efficiency	Improvement/Modification	Recyclability	References
CdS-TiO_2	Hydrothermal synthesis	Sunlight	Methylene blue/77% in 180 min	Extended light absorption to visible range Reduced recombination	–	Gomathi Thanga Keerthana and Murugakoothan [40]
CdS-TiO_2	Hydrothermal synthesis	Visible light	Rhodamine B/greater than 80%	Enhanced charge separation Extended light absorption to visible range	–	Li et al. [66]
C_3N_4/TiO_2	Microwave-assisted synthesis	Visible light	Reactive red 4/rate constant 5.3 times higher than P25	Visible light photocatalysis Reduced recombination	–	Azami et al. [9]

[38, 103, 123]. Some of these polymers can also enhance the visible light activity of the composite. A polymer, poly hydroxy butyrate, was used as a substrate for immobilization with solar light activity. The authors reported biodegradability of the polymer, but that can lead to the release of TiO_2 nanoparticles to the environment which is not desirable [139].

TiO_2 was successfully immobilized on clay with improved efficiency owing to adsorption and concentration of contaminants near to the photocatalyst [119, 129]. Visible light activity was also reported with hybrid kaolin clay substrate [129].

Many natural materials like chitosan, 3D eggshell membrane, xylan, cellulose, etc., are successfully used in the immobilization of TiO_2 [36, 38, 42, 68, 73, 142, 144]. These membranes concentrate the dye near TiO_2 by adsorption, improves the recovery of the photocatalyst and reduce agglomeration. The possibility of photocatalytic degradation of the cellulose membranes needs to considered [123]. Another issue of the cellulose membrane is its negative charge due to which retention aids need to be incorporated in the photocatalyst for effective degradation of anionic dye [98] (Table 5).

5.6 *Co-modified Composites*

Of late co-modified composites with more than one additive are gaining importance. The synergistic action of the multiple additives aid in overcoming multiple limitations of the TiO_2 photocatalyst. Brindha and Sivakumar [17] developed nitrogen and sulphur co-doped TiO_2/rGO composite, where rGO acted as acceptor and transporter of photogenerated electron thereby reducing recombination and anion doping reduced the band gap energy enabling visible light photocatalysis. There was a considerable effect of the graphene dose and 5%-rGO was found to be optimal [17]. The efficacy of TiO_2 composites with N-doped graphene and B-doped graphene was studied by Gopalakrishnan et al. [41]. It was found that the type of dopant affects the behaviour of the photocatalyst. During degradation of methylene blue and rhodamine B using these photocatalysts, it was observed that boron-doped graphene/TiO_2 performed best in degrading methylene blue and nitrogen-doped graphene/TiO_2 performed well in the removal of rhodamine B [41]. TiO_2/B-nanowire/rGO was reported to result in band gap engineering and had visible light activity, increased surface area and adsorption and reduced recombination [76]. Nitrogen and palladium co-doped TiO_2 embedded in polysulfone membrane could effectively be used in membrane photocatalytic reactor with visible light activity [60].

A comparative assessment of semiconductor composite and first-row transition metal-doped semiconductor composite indicated that the metal doping has improved the efficiency of the photocatalyst and enhanced light absorption. The efficiency of the different composites follows the order Ni-SiO_2/TiO_2 > Cr-SiO_2/TiO_2 > Zn-SiO_2/TiO_2 > Co-SiO_2/TiO_2 > Cu-SiO_2/TiO_2 > SiO_2/TiO_2 [2]. Another composite of metal-doped TiO_2 is Ni-TiO_2/GO [95]. Boikanyo et al. [16] synthesized erbium and MWCNT modified TiO_2 immobilized on a biopolymer. Cellulose acetate and

Table 5 Performance of support structure-based TiO_2 composites

Catalyst	Preparation procedure	Light source	Dye/efficiency	Improvement/Modification	Recyclability	References
TiO_2/polyaniline	In-situ chemical oxidative polymerization in the presence of hydrothermally synthesized TiO_2 nanoparticles	UV light	Rhodamine B/approximately 80% Methylene blue/approximately 70%	Improved photocatalytic activity due to sensitization Mass transfer limitation addressed Reduced recombination	Considerable reduction in efficiency after 3 cycles	Reddy et al. [103]
Boron nitride-TiO_2	Ice bath method	Visible light	Methylene Blue/79%	Mass transfer limitation addressed Prevent aggregation Visible light photocatalysis	–	Singh et al. [115]
TiO_2 coated on transparent glass, glazed ceramic tile, and stainless steel	TiO_2 prepared by sol–gel method coated on substrates	UV light	Methylene blue/more than 90% for TiO_2 coated glass and ceramic tile approximately 40% for TiO_2 coated stainless steel	Easy separation from the reactor Well distributed catalyst	No reduction in efficiency after 20 cycles for TiO_2 coated glass	Sirirerkratana et al. [116]
Microfibrous palygorskite clay/TiO_2 nanocomposite film on glass substrate	Acetic acid-based sol–gel route followed by thermal treatment	UV light	Basic blue 41/more than 90% for photocatalyst with ratio of palygorskite clay/TiO_2 as 3:2	Easy separation from the reactor Reduced agglomeration Mass transfer limitation addressed	Less than 10% reduction in efficiency after 3 cycles	Stathatos et al. [119]

(continued)

Table 5 (continued)

Catalyst	Preparation procedure	Light source	Dye/efficiency	Improvement/Modification	Recyclability	References
poly(vinylidene fluoride)–anatase TiO_2	Phase inversion technique	UV light	Brilliant green Indigo carmin/complete decolourization on prolonged exposure. Not as efficient as pure TiO_2	Mass transfer limitation addressed Reduced agglomeration	–	Tahiri Alaoui et al. [123]
Hybrid kaolin/TiO_2	Two-step wet precipitation method	Visible light	methylene blue/90%	Visible light photocatalysis Mass transport limitation addressed	–	Wongso et al. [129]
Polyhydroxybutyrate/TiO_2	Solvent-cast technique	Solar light	Methylene Blue/96%	Easy separation from the reactor Solar light photocatalysis	–	Yew et al. [139]
Poly(3-hexylthiophene)/TiO_2	polymer/inorganic blending technique	UV and visible light	Methyl orange/more than 80%	Visible light photocatalysis Reduced recombination Mass transfer limitation addressed	–	Zhu and Dan [145]
Polythiophene/TiO_2	In situ chemical oxidative polymerization	UV and visible light	Methyl orange/more than 90% in UV light and more than 80% in visible light	Visible light photocatalysis Mass transfer limitation addressed	–	Zhu et al. [146]
Xylan/Poly Vinyl Acetate/TiO_2	Precipitation method	Visible light	Ethyl violet, Astrazon Brilliant Red 4G/more than 90%	Visible light photocatalysis Mass transfer limitation addressed	A negligible reduction in efficiency after 3 cycles	Liu et al. [73]

(continued)

Table 5 (continued)

Catalyst	Preparation procedure	Light source	Dye/efficiency	Improvement/Modification	Recyclability	References
3D eggshell membrane-modified with polyethyleneimine supported anatase TiO_2	Self-assembly method	Simulated sunlight	Rhodamine B/more than 60%	Mass transfer limitation addressed Prevent aggregation	–	Li et al. [68]
Micro fibrillated cellulose—polyamide-amine epichlorohydrin—TiO_2 nanoparticles	Two-step mixing process	UV light	Methyl orange/more than 80%	Mass transfer limitation addressed Prevent aggregation Easy separation from the reactor	Negligible reduction in efficiency after 3 cycles	Garusinghe et al. [38]
Natural cellulosic *Juncus effusus* fiber/TiO_2 composite	Sol–gel	Simulated sunlight	C.I. Reactive Red 120, C.I. Direct Yellow 12, and methylene blue/100% (approximate)	Mass transfer limitation addressed Easy separation from the reactor	Negligible reduction in efficiency after 5 cycles	Zhou et al. [147]

polycaprol acetone were the biopolymers attempted. In this composite, Er^{3+} acts as a photosensitizer and aids in red shift by band gap engineering. MWCNT aids in reducing recombination and concentrating the contaminant near the photocatalyst and immobilization on biopolymer helps in easy recovery of the photocatalyst from the reactor. Overall, this combination addresses the major issues posed in TiO_2 photocatalysis.

Co-modification of the semiconductor composite ZnO/TiO_2 with different weight percentage of rGO indicated that 5% rGO aided in efficiency enhancement of the photocatalyst. The improvement in overall efficiency was due to adsorption and concentration of the dye by rGO and the effective transport of photogenerated electrons to rGO, thereby reducing recombination [86]. Substantial increase in photodegradation efficiency was obtained by a similar mechanism in $TiO_2/ZnO/Bi_2O_3$-graphene nanofibre composite and ZnO-graphene-TiO_2 composite [52, 88]. ZnO-graphene-TiO_2 composite was 3 times more efficient than pure TiO_2 [88] Another co-modification of semiconductor composite is Cu_2O/TiO_2 composite immobilized with aluminosilicate inorganic polymer [33] and CdS-graphene-TiO_2 co-modified composite. Performance of co-modified composites is depicted in Table 6.

6 Discussion and Conclusions

Photocatalysis is a potential technology for the treatment of dye-containing wastewater and TiO_2 is the most widely studied photocatalyst. However, pure TiO_2, even though efficient has many limitations, which restrict its field application. Many TiO_2 composites are developed to address the shortfall of pure TiO_2 photocatalysis. The studies indicate the enhanced activity of the catalysts. Co-modified composites are having good prospects for field application as they can address multiple limitations of pure TiO_2 photocatalysis. However, the studies are mostly carried out under simulated conditions and only limited testing is done under real conditions. In a few studies on commercial dyes, a significant reduction in efficiency in comparison to other dyes was observed. Hence it is difficult to assess the applicability in the field. Also, the studies on composites are limited to the development of novel composites and testing. The influence of modification on pH_{zpc} and the effect of operational parameters are studied rarely. Hence, the novel catalysts developed may be good for a group of dyes and not good for other types of dyes.

Only very few studies are focused on reactor design and field testing. This indicates that there is a long way to go for the field implementation of photocatalytic treatment of dye wastewater using TiO_2 and its composites.

Table 6 Preparation procedure and performance of co-modified composites

Catalyst	Preparation procedure	Light source	Dye/efficiency	Improvement/Modification	Recyclability	References
N, S co-doped TiO_2/graphene	Single-step hydrothermal method	Visible light	Congo red, Methylene blue, reactive orange 16/more than 95%	High surface area Improved visible light absorption Reduce recombination	No significant reduction in efficiency even after 4 cycles	Brindha and Sivakumar [17]
B-doped graphene TiO_2/and N-doped graphene TiO_2	Arc discharge method	Mercury lamp	Methylene blue/more than 90% with B-doped graphene TiO_2 and more than70% with N-doped graphene TiO_2 Rhodamine B/more than 60% with N-doped graphene TiO_2 and more than 30% with B-doped graphene TiO_2	Improved visible light absorption Mass transport limitation addressed	–	Gopalakrishnan et al. [41]

(continued)

Table 6 (continued)

Catalyst	Preparation procedure	Light source	Dye/efficiency	Improvement/Modification	Recyclability	References
Reduced graphene oxide laminated TiO_2-B nanowire composites	High pressure and high-temperature hydrothermal method	Visible light	Rhodamine B/more than 80%	Improved visible light absorption Mass transport limitation addressed Reduce recombination	No significant reduction in efficiency even after 4 cycles	Makal and Das [76]
N, Pd co-doped TiO_2 were embedded in a polysulfone polymeric membrane	Phase inversion method	Visible light	Eosin yellow/more than 90%	Visible light photocatalysis Easy recovery of photocatalyst from the reactor Reduce recombination	–	Kuvarega et al. [60]
Ni-incorporated TiO_2/GO	Microwave-assisted one-pot method	Visible light	Rhodamine B/50% approximately	Visible light photocatalysis Reduce recombination Mass transport limitation addressed	–	Pham et al. [95]
Er-MWCNT-TiO_2 immobilized on biopolymer	Wet impregnation Co-precipitation	Visible light	Reactive red 120	Visible light photocatalysis Easy recovery of photocatalyst from the reactor Mass transport limitation addressed	–	Boikanyo et al. [16]
ZnO-graphene-TiO_2	Solvothermal process	Visible light	Rhodamine B (RhB)/80%	Visible light photocatalysis Reduce recombination Mass transport limitation addressed	No significant reduction in efficiency even after 4 cycles	Nuengmatcha et al. [88]
Cu_2O/TiO_2 with aluminosilicate inorganic polymer	Alcohol-aqueous-based chemical precipitation method	UV light	Methylene blue/greater than 95%	Mass transport limitation addressed	–	Falah and MacKenzie [33]

References

1. Adly MS, El-Dafrawy SM, El-Hakam SA (2019) Application of nanostructured graphene oxide/titanium dioxide composites for photocatalytic degradation of rhodamine B and acid green 25 dyes. J Mater Res Technol 8:5610–5622. https://doi.org/10.1016/j.jmrt.2019.09.029
2. Ahmad S, Saeed A (2019) Synthesis of metal/silica/titania composites for the photocatalytic removal of methylene blue dye. J Chem 2019:1–6. https://doi.org/10.1155/2019/9010289
3. Alamelu K, Raja V, Shiamala L, Jaffar Ali BM (2018) Biphasic TiO_2 nanoparticles decorated graphene nanosheets for visible light driven photocatalytic degradation of organic dyes. Appl Surf Sci 430:145–154. https://doi.org/10.1016/j.apsusc.2017.05.054
4. Ali I, Kim J-O (2018) Visible-light-assisted photocatalytic activity of bismuth-TiO_2 nanotube composites for chromium reduction and dye degradation. Chemosphere 207:285–292. https://doi.org/10.1016/j.chemosphere.2018.05.075
5. Ambrus Z, Balázs N, Alapi T et al (2008) Synthesis, structure and photocatalytic properties of Fe(III)-doped TiO_2 prepared from $TiCl_3$. Appl Catal B Environ 81:27–37. https://doi.org/10.1016/j.apcatb.2007.11.041
6. Andriantsiferana C, Mohamed EF, Delmas H (2014) Photocatalytic degradation of an azo-dye on TiO_2/activated carbon composite material. Environ Technol 35:355–363. https://doi.org/10.1080/09593330.2013.828094
7. Asiltürk M, Sayılkan F, Arpaç E (2009) Effect of Fe^{3+} ion doping to TiO_2 on the photocatalytic degradation of Malachite Green dye under UV and vis-irradiation. J Photochem Photobiol A Chem 203:64–71. https://doi.org/10.1016/j.jphotochem.2008.12.021
8. Atchudan R, Jebakumar Immanuel Edison TN, Perumal S et al (2017) Effective photocatalytic degradation of anthropogenic dyes using graphene oxide grafting titanium dioxide nanoparticles under UV-light irradiation. J Photochem Photobiol A Chem 333:92–104. https://doi.org/10.1016/j.jphotochem.2016.10.021
9. Azami MS, Ismail K, Ishak MAM et al (2020) Formation of an amorphous carbon nitride/titania composite for photocatalytic degradation of RR4 dye. J Water Process Eng 35: https://doi.org/10.1016/j.jwpe.2020.101209
10. Badawy MI, Ali MEM, Ghaly MY, El-Missiry MA (2015) Mesoporous simonkolleite–TiO_2 nanostructured composite for simultaneous photocatalytic hydrogen production and dye decontamination. Process Saf Environ Prot 94:11–17. https://doi.org/10.1016/j.psep.2014.12.001
11. Bai S, Li H, Guan Y, Jiang S (2011) The enhanced photocatalytic activity of CdS/TiO_2 nanocomposites by controlling CdS dispersion on TiO_2 nanotubes. Appl Surf Sci 257:6406–6409. https://doi.org/10.1016/j.apsusc.2011.02.007
12. Bai S, Liu H, Sun J et al (2015) Improvement of TiO_2 photocatalytic properties under visible light by WO_3/TiO_2 and MoO_3/TiO_2 composites. Appl Surf Sci 338:61–68. https://doi.org/10.1016/j.apsusc.2015.02.103
13. Bailón-García E, Elmouwahidi A, Álvarez MA et al (2017) New carbon xerogel-TiO_2 composites with high performance as visible-light photocatalysts for dye mineralization. Appl Catal B Environ 201:29–40. https://doi.org/10.1016/j.apcatb.2016.08.015
14. Bento RT, Correa OV, Pillis MF (2019) Photocatalytic activity of undoped and sulfur-doped TiO_2 films grown by MOCVD for water treatment under visible light. J Eur Ceram Soc 39:3498–3504. https://doi.org/10.1016/j.jeurceramsoc.2019.02.046
15. Bessekhouad Y, Robert D, Weber J-V, Chaoui N (2004) Effect of alkaline-doped TiO_2 on photocatalytic efficiency. J Photochem Photobiol A Chem 167:49–57. https://doi.org/10.1016/j.jphotochem.2003.12.001
16. Boikanyo D, Mishra SB, Nxumalo EN et al (2019) Erbium and MWCNT-modified titanium dioxide nanocomposites for the photocatalytic degradation of Azo dyes
17. Brindha A, Sivakumar T (2017) Visible active N, S co-doped TiO_2/graphene photocatalysts for the degradation of hazardous dyes. J Photochem Photobiol A Chem 340:146–156. https://doi.org/10.1016/j.jphotochem.2017.03.010

18. Byrne C, Subramanian G, Pillai SC (2018) Recent advances in photocatalysis for environmental applications. J Environ Chem Eng 6:3531–3555. https://doi.org/10.1016/j.jece.2017.07.080
19. Cantarella M, Sanz R, Buccheri MA et al (2016) Immobilization of nanomaterials in PMMA composites for photocatalytic removal of dyes, phenols and bacteria from water. J Photochem Photobiol A Chem 321:1–11. https://doi.org/10.1016/j.jphotochem.2016.01.020
20. Chan SHS, Yeong WuT, Juan JC, Teh CY (2011) Recent developments of metal oxide semiconductors as photocatalysts in advanced oxidation processes (AOPs) for treatment of dye waste-water. J Chem Technol Biotechnol 86:1130–1158. https://doi.org/10.1002/jctb.2636
21. Chen D, Jiang Z, Geng J et al (2007) Carbon and nitrogen co-doped TiO_2 with enhanced visible-light photocatalytic activity. Ind Eng Chem Res 46:2741–2746. https://doi.org/10.1021/ie061491k
22. Chen L-C, Huang C-M, Tsai F-R (2007) Characterization and photocatalytic activity of K+-doped TiO_2 photocatalysts. J Mol Catal A: Chem 265:133–140. https://doi.org/10.1016/j.molcata.2006.10.011
23. Chen Q, Jiang D, Shi W et al (2009) Visible-light-activated Ce–Si co-doped TiO_2 photocatalyst. Appl Surf Sci 255:7918–7924. https://doi.org/10.1016/j.apsusc.2009.04.167
24. Chen J, Qian Y, Wei X (2010) Comparison of magnetic-nanometer titanium dioxide/ferriferous oxide (TiO_2/Fe_3O_4) composite photocatalyst prepared by acid–sol and homogeneous precipitation methods. J Mater Sci 45:6018–6024. https://doi.org/10.1007/s10853-010-4685-z
25. Cheng X, Yu X, Xing Z, Wan J (2012) Enhanced photocatalytic activity of nitrogen doped TiO_2 anatase nano-particle under simulated sunlight irradiation. Energy Procedia 16:598–605. https://doi.org/10.1016/j.egypro.2012.01.096
26. Couselo N, García Einschlag FS, Candal RJ, Jobbágy M (2008) Tungsten-doped TiO_2 vs. pure TiO_2 photocatalysts: effects on photobleaching kinetics and mechanism. J Phys Chem C 112:1094–1100. https://doi.org/10.1021/jp0769781
27. Da Dalt S, Alves AK, Bergmann CP (2013) Photocatalytic degradation of methyl orange dye in water solutions in the presence of MWCNT/TiO_2 composites. Mater Res Bull 48:1845–1850. https://doi.org/10.1016/j.materresbull.2013.01.022
28. Dahl M, Liu Y, Yin Y (2014) Composite titanium dioxide nanomaterials. Chem Rev 114:9853–9889. https://doi.org/10.1021/cr400634p
29. Debnath S, Ballav N, Nyoni H et al (2015) Optimization and mechanism elucidation of the catalytic photo-degradation of the dyes Eosin Yellow (EY) and Naphthol blue black (NBB) by a polyaniline-coated titanium dioxide nanocomposite. Appl Catal B Environ 163:330–342. https://doi.org/10.1016/j.apcatb.2014.08.011
30. Demirci S, Dikici T, Yurddaskal M et al (2016) Synthesis and characterization of Ag doped TiO_2 heterojunction films and their photocatalytic performances. Appl Surf Sci 390:591–601. https://doi.org/10.1016/j.apsusc.2016.08.145
31. Duta A, Visa M (2015) Simultaneous removal of two industrial dyes by adsorption and photocatalysis on a fly-ash–TiO_2 composite. J Photochem Photobiol A Chem 306:21–30. https://doi.org/10.1016/j.jphotochem.2015.03.007
32. Elghniji K, Atyaoui A, Livraghi S et al (2012) Synthesis and characterization of Fe^{3+} doped TiO_2 nanoparticles and films and their performance for photocurrent response under UV illumination. J Alloys Compd 541:421–427. https://doi.org/10.1016/j.jallcom.2012.07.010
33. Falah M, MacKenzie KJD (2015) Synthesis and properties of novel photoactive composites of P25 titanium dioxide and copper (I) oxide with inorganic polymers. Ceram Int 41:13702–13708. https://doi.org/10.1016/j.ceramint.2015.07.198
34. Fang W, Xing M, Zhang J (2017) Modifications on reduced titanium dioxide photocatalysts: a review. J Photochem Photobiol C Photochem Rev 32:21–39. https://doi.org/10.1016/j.jphotochemrev.2017.05.003
35. Fang T, Chen X, Wang M et al (2020) Cuprous oxide/titanium dioxide composite photocatalytic decolorization of reactive brilliant red X-3B dyes wastewater under visible light. Res Chem Intermed 46:5459–5477. https://doi.org/10.1007/s11164-020-04272-y

36. Farzana MH, Meenakshi S (2014) Synergistic effect of chitosan and titanium dioxide on the removal of toxic dyes by the photodegradation technique. Ind Eng Chem Res 53:55–63. https://doi.org/10.1021/ie402347g
37. Feng L, van Hullebusch ED, Rodrigo MA et al (2013) Removal of residual anti-inflammatory and analgesic pharmaceuticals from aqueous systems by electrochemical advanced oxidation processes. A review. Chem Eng J 228:944–964. https://doi.org/10.1016/j.cej.2013.05.061
38. Garusinghe UM, Raghuwanshi VS, Batchelor W, Garnier G (2018) Water resistant cellulose—titanium dioxide composites for photocatalysis. Sci Rep 8:2306. https://doi.org/10.1038/s41598-018-20569-w
39. Gayathri S, Kottaisamy M, Ramakrishnan V (2015) Facile microwave-assisted synthesis of titanium dioxide decorated graphene nanocomposite for photodegradation of organic dyes. AIP Adv 5: https://doi.org/10.1063/1.4938544
40. Gomathi Thanga Keerthana B, Murugakoothan P (2019) Synthesis and characterization of CdS/TiO_2 nanocomposite: methylene blue adsorption and enhanced photocatalytic activities. Vacuum 159:476–481. https://doi.org/10.1016/j.vacuum.2018.10.082
41. Gopalakrishnan K, Joshi HM, Kumar P et al (2011) Selectivity in the photocatalytic properties of the composites of TiO_2 nanoparticles with B- and N-doped graphenes. Chem Phys Lett 511:304–308. https://doi.org/10.1016/j.cplett.2011.06.033
42. Habiba U, Islam MS, Siddique TA et al (2016) Adsorption and photocatalytic degradation of anionic dyes on Chitosan/PVA/Na–Titanate/TiO_2 composites synthesized by solution casting method. Carbohydr Polym 149:317–331. https://doi.org/10.1016/j.carbpol.2016.04.127
43. Habibi MH, Esfahani MN, Egerton TA (2007) Photochemical characterization and photocatalytic properties of a nanostructure composite TiO_2 film. Int J Photoenergy 2007:1–8. https://doi.org/10.1155/2007/13653
44. Han F, Kambala VSR, Srinivasan M et al (2009) Tailored titanium dioxide photocatalysts for the degradation of organic dyes in wastewater treatment: a review. Appl Catal A Gen 359:25–40. https://doi.org/10.1016/j.apcata.2009.02.043
45. Hashimoto K, Irie H, Fujishima A (2005) TiO_2 photocatalysis: a historical overview and future prospects. Jpn J Appl Phys 44:8269–8285. https://doi.org/10.1143/JJAP.44.8269
46. Hickman R, Walker E, Chowdhury S (2018) TiO_2-PDMS composite sponge for adsorption and solar mediated photodegradation of dye pollutants. J Water Process Eng 24:74–82. https://doi.org/10.1016/j.jwpe.2018.05.015
47. Higashimoto S (2019) Titanium-dioxide-based visible-light-sensitive photocatalysis: mechanistic insight and applications. Catalysts 9:201. https://doi.org/10.3390/catal9020201
48. Hurum DC, Agrios AG, Gray KA et al (2003) Explaining the enhanced photocatalytic activity of Degussa P25 mixed-phase TiO_2 using EPR. J Phys Chem B 107:4545–4549. https://doi.org/10.1021/jp0273934
49. Ibukun O, Evans PE, Dowben PA, Kyung Jeong H (2019) Titanium dioxide-molybdenum disulfide for photocatalytic degradation of methylene blue. Chem Phys 525: https://doi.org/10.1016/j.chemphys.2019.110419
50. Jiang T, Zhang L, Ji M et al (2013) Carbon nanotubes/TiO_2 nanotubes composite photocatalysts for efficient degradation of methyl orange dye. Particuology 11:737–742. https://doi.org/10.1016/j.partic.2012.07.008
51. Kangwansupamonkon W, Jitbunpot W, Kiatkamjornwong S (2010) Photocatalytic efficiency of TiO_2/poly[acrylamide-co-(acrylic acid)] composite for textile dye degradation. Polym Degrad Stab 95:1894–1902. https://doi.org/10.1016/j.polymdegradstab.2010.04.019
52. Kanjwal MA, Lo KKS, Leung WW-F (2019) Graphene composite nanofibers as a high-performance photocatalyst for environmental remediation. Sep Purif Technol 215:602–611. https://doi.org/10.1016/j.seppur.2019.01.044
53. Karimi L (2017) Combination of mesoporous titanium dioxide with MoS_2 nanosheets for high photocatalytic activity. Polish J Chem Technol 19:56–60. https://doi.org/10.1515/pjct-2017-0028
54. Kedem S, Schmidt J, Paz Y, Cohen Y (2005) Composite polymer nanofibers with carbon nanotubes and titanium dioxide particles. Langmuir 21:5600–5604. https://doi.org/10.1021/la0502443

55. Kianfar AH, Dehghani P, Momeni MM (2016) Photo-catalytic degradation of methylene blue over nano titanium/nickel oxide prepared from supported Schiff base complex on titanium dioxide. J Mater Sci: Mater Electron 27:3368–3375. https://doi.org/10.1007/s10854-015-4167-9
56. Kim S, Hwang S-J, Choi W (2005) Visible light active platinum-ion-doped TiO_2 photocatalyst. J Phys Chem B 109:24260–24267. https://doi.org/10.1021/jp055278y
57. Koohestani H, Sadrnezhaad SK (2016) Photocatalytic degradation of methyl orange and cyanide by using TiO_2/CuO composite. Desalin Water Treat 57:22029–22038. https://doi.org/10.1080/19443994.2015.1132395
58. Kratofil Krehula L, Stjepanović J, Perlog M et al (2019) Conducting polymer polypyrrole and titanium dioxide nanocomposites for photocatalysis of RR45 dye under visible light. Polym Bull 76:1697–1715. https://doi.org/10.1007/s00289-018-2463-2
59. Kutuzova AS, Dontsova TA (2019) Characterization and properties of TiO_2–SnO_2 nanocomposites, obtained by hydrolysis method. Appl Nanosci 9:873–880. https://doi.org/10.1007/s13204-018-0754-4
60. Kuvarega AT, Khumalo N, Dlamini D, Mamba BB (2018) Polysulfone/N, Pd co-doped TiO_2 composite membranes for photocatalytic dye degradation. Sep Purif Technol 191:122–133. https://doi.org/10.1016/j.seppur.2017.07.064
61. Lai YK, Huang JY, Zhang HF et al (2010) Nitrogen-doped TiO_2 nanotube array films with enhanced photocatalytic activity under various light sources. J Hazard Mater 184:855–863. https://doi.org/10.1016/j.jhazmat.2010.08.121
62. Laysandra L, Sari MWMK, Soetaredjo FE et al (2017) Adsorption and photocatalytic performance of bentonite-titanium dioxide composites for methylene blue and rhodamine B decoloration. Heliyon 3: https://doi.org/10.1016/j.heliyon.2017.e00488
63. Lazar M, Varghese S, Nair S (2012) Photocatalytic water treatment by titanium dioxide: recent updates. Catalysts 2:572–601. https://doi.org/10.3390/catal2040572
64. Le P, Hieu L, Lam T-N et al (2018) Enhanced photocatalytic performance of nitrogen-doped TiO_2 nanotube arrays using a simple annealing process. Micromachines 9:618. https://doi.org/10.3390/mi9120618
65. Lee YC, Hong YP, Lee HY et al (2003) Photocatalysis and hydrophilicity of doped TiO_2 thin films. J Colloid Interface Sci 267:127–131. https://doi.org/10.1016/S0021-9797(03)00603-9
66. Li W, Cui X, Wang P et al (2013) Enhanced photosensitized degradation of rhodamine B on CdS/TiO_2 nanocomposites under visible light irradiation. Mater Res Bull 48:3025–3031. https://doi.org/10.1016/j.materresbull.2013.04.057
67. Li H, Xing J, Xia Z, Chen J (2014) Preparation of extremely smooth and boron-fluorine co-doped TiO_2 nanotube arrays with enhanced photoelectrochemical and photocatalytic performance. Electrochim Acta 139:331–336. https://doi.org/10.1016/j.electacta.2014.06.172
68. Li Y, Zhou J, Fan Y et al (2017) Preparation of environment-friendly 3D eggshell membrane-supported anatase TiO_2 as a reusable photocatalyst for degradation of organic dyes. Chem Phys Lett 689:142–147. https://doi.org/10.1016/j.cplett.2017.10.019
69. Liqiang J, Xiaojun S, Baifu X et al (2004) The preparation and characterization of La doped TiO_2 nanoparticles and their photocatalytic activity. J Solid State Chem 177:3375–3382. https://doi.org/10.1016/j.jssc.2004.05.064
70. Liu Y, Chen X, Li J, Burda C (2005) Photocatalytic degradation of azo dyes by nitrogen-doped TiO_2 nanocatalysts. Chemosphere 61:11–18. https://doi.org/10.1016/j.chemosphere.2005.03.069
71. Liu S, Sun H, Liu S, Wang S (2013) Graphene facilitated visible light photodegradation of methylene blue over titanium dioxide photocatalysts. Chem Eng J 214:298–303. https://doi.org/10.1016/j.cej.2012.10.058
72. Liu H, Guo W, Li Y et al (2018) Photocatalytic degradation of sixteen organic dyes by TiO_2/WO_3-coated magnetic nanoparticles under simulated visible light and solar light. J Environ Chem Eng 6:59–67. https://doi.org/10.1016/j.jece.2017.11.063
73. Liu Z, Liu R, Yi Y et al (2019) Photocatalytic degradation of dyes over a xylan/PVA/TiO_2 composite under visible light irradiation. Carbohydr Polym 223: https://doi.org/10.1016/j.carbpol.2019.115081

74. Magalhães F, Moura FCC, Lago RM (2011) TiO_2/LDPE composites: a new floating photocatalyst for solar degradation of organic contaminants. Desalination 276:266–271. https://doi.org/10.1016/j.desal.2011.03.061
75. Mahadik MA, An GW, David S et al (2017) Fabrication of A/R-TiO_2 composite for enhanced photoelectrochemical performance: solar hydrogen generation and dye degradation. Appl Surf Sci 426:833–843. https://doi.org/10.1016/j.apsusc.2017.07.179
76. Makal P, Das D (2019) Superior photocatalytic dye degradation under visible light by reduced graphene oxide laminated TiO_2-B nanowire composite. J Environ Chem Eng 7: https://doi.org/10.1016/j.jece.2019.103358
77. Matos J, García A, Zhao L, Titirici MM (2010) Solvothermal carbon-doped TiO_2 photocatalyst for the enhanced methylene blue degradation under visible light. Appl Catal A Gen 390:175–182. https://doi.org/10.1016/j.apcata.2010.10.009
78. Mohamed A, El-Sayed R, Osman TA et al (2016) Composite nanofibers for highly efficient photocatalytic degradation of organic dyes from contaminated water. Environ Res 145:18–25. https://doi.org/10.1016/j.envres.2015.09.024
79. Momeni MM, Ghayeb Y (2016) Preparation of cobalt coated TiO_2 and WO_3–TiO_2 nanotube films via photo-assisted deposition with enhanced photocatalytic activity under visible light illumination. Ceram Int 42:7014–7022. https://doi.org/10.1016/j.ceramint.2016.01.089
80. Mukthar Ali M, Sandhya KY (2016) Selective photodegradation and enhanced photo electrochemical properties of titanium dioxide–graphene composite with exposed (001) facets made by photochemical method. Sol Energy Mater Sol Cells 144:748–757. https://doi.org/10.1016/j.solmat.2015.10.025
81. Muthirulan P, Devi CN, Sundaram MM (2014) TiO_2 wrapped graphene as a high performance photocatalyst for acid orange 7 dye degradation under solar/UV light irradiations. Ceram Int 40:5945–5957. https://doi.org/10.1016/j.ceramint.2013.11.042
82. Muthirulan P, Nirmala Devi C, Meenakshi Sundaram M (2017) Synchronous role of coupled adsorption and photocatalytic degradation on CAC–TiO_2 composite generating excellent mineralization of alizarin cyanine green dye in aqueous solution. Arab J Chem 10:S1477–S1483. https://doi.org/10.1016/j.arabjc.2013.04.028
83. Nam S-H, Kim TK, Boo J-H (2012) Physical property and photo-catalytic activity of sulfur doped TiO_2 catalysts responding to visible light. Catal Today 185:259–262. https://doi.org/10.1016/j.cattod.2011.07.033
84. Nasr M, Eid C, Habchi R et al (2018) Recent progress on titanium dioxide nanomaterials for photocatalytic applications. Chemsuschem 11:3023–3047. https://doi.org/10.1002/cssc.201800874
85. Nguyen-Phan T-D, Pham VH, Shin EW et al (2011) The role of graphene oxide content on the adsorption-enhanced photocatalysis of titanium dioxide/graphene oxide composites. Chem Eng J 170:226–232. https://doi.org/10.1016/j.cej.2011.03.060
86. Nguyen CH, Juang R-S (2019) Efficient removal of methylene blue dye by a hybrid adsorption–photocatalysis process using reduced graphene oxide/titanate nanotube composites for water reuse. J Ind Eng Chem 76:296–309. https://doi.org/10.1016/j.jiec.2019.03.054
87. Nguyen CH, Fu C-C, Juang R-S (2018) Degradation of methylene blue and methyl orange by palladium-doped TiO_2 photocatalysis for water reuse: efficiency and degradation pathways. J Clean Prod 202:413–427. https://doi.org/10.1016/j.jclepro.2018.08.110
88. Nuengmatcha P, Chanthai S, Mahachai R, Oh W-C (2016) Visible light-driven photocatalytic degradation of rhodamine B and industrial dyes (texbrite BAC-L and texbrite NFW-L) by ZnO-graphene-TiO2 composite. J Environ Chem Eng 4:2170–2177. https://doi.org/10.1016/j.jece.2016.03.045
89. Oh WC, Jung AR, Ko WB (2009) Characterization and relative photonic efficiencies of a new nanocarbon/TiO_2 composite photocatalyst designed for organic dye decomposition and bactericidal activity. Mater Sci Eng C 29:1338–1347. https://doi.org/10.1016/j.msec.2008.10.034
90. Ohno T, Akiyoshi M, Umebayashi T et al (2004) Preparation of S-doped TiO_2 photocatalysts and their photocatalytic activities under visible light. Appl Catal A Gen 265:115–121. https://doi.org/10.1016/j.apcata.2004.01.007

91. Pal M, Bera S, Jana S (2015) Sol-gel based simonkolleite nanopetals with SnO_2 nanoparticles in graphite-like amorphous carbon as an efficient and reusable photocatalyst. RSC Adv 5:75062–75074. https://doi.org/10.1039/c5ra12322d
92. Park CY, Kefayat U, Vikram N et al (2013) Preparation of novel CdS-graphene/TiO_2 composites with high photocatalytic activity for methylene blue dye under visible light. Bull Mater Sci 36:869–876. https://doi.org/10.1007/s12034-013-0533-5
93. Pastrana-Martínez LM, Morales-Torres S, Likodimos V et al (2012) Advanced nanostructured photocatalysts based on reduced graphene oxide–TiO_2 composites for degradation of diphenhydramine pharmaceutical and methyl orange dye. Appl Catal B Environ 123–124:241–256. https://doi.org/10.1016/j.apcatb.2012.04.045
94. Pelaez M, Nolan NT, Pillai SC et al (2012) A review on the visible light active titanium dioxide photocatalysts for environmental applications. Appl Catal B Environ 125:331–349. https://doi.org/10.1016/j.apcatb.2012.05.036
95. Pham T-T, Nguyen-Huy C, Shin EW (2016) Facile one-pot synthesis of nickel-incorporated titanium dioxide/graphene oxide composites: enhancement of photodegradation under visible-irradiation. Appl Surf Sci 377:301–310. https://doi.org/10.1016/j.apsusc.2016.03.144
96. Pillai IMS, Gupta AK (2016) Electrochemical degradation of malachite green: Multiparameter optimization, pathway identification and toxicity analysis. J Environ Sci Heal Part A Environ Sci Eng Toxicol 51:1091–1099
97. Pillai IMS, Gupta AK, Tiwari MK (2015) Multivariate optimization for electrochemical oxidation of methyl orange: pathway identification and toxicity analysis. J Environ Sci Health A Tox Hazard Subst Environ Eng 50:301–310. https://doi.org/10.1080/10934529.2015.981119
98. Raghuwanshi VS, Garusinghe UM, Batchelor W, Garnier G (2020) Polyamide-amine-epichlorohydrin (PAE) induced TiO_2 nanoparticles assembly in cellulose network. J Colloid Interface Sci 575:317–325. https://doi.org/10.1016/j.jcis.2020.04.121
99. Rajamanickam D, Shanthi M (2014) Photocatalytic degradation of an azo dye Sunset Yellow under UV-A light using TiO_2/CAC composite catalysts. Spectrochim Acta Part A Mol Biomol Spectrosc 128:100–108. https://doi.org/10.1016/j.saa.2014.02.126
100. Rajamanickam D, Dhatshanamurthi P, Shanthi M (2015) Preparation and characterization of SeO_2/TiO_2 composite photocatalyst with excellent performance for sunset yellow azo dye degradation under natural sunlight illumination. Spectrochim Acta Part A Mol Biomol Spectrosc 138:489–498. https://doi.org/10.1016/j.saa.2014.11.070
101. Rajkumar K, Vairaselvi P, Saravanan P et al (2015) Visible-light-driven SnO_2/TiO_2 nanotube nanocomposite for textile effluent degradation. RSC Adv 5:20424–20431. https://doi.org/10.1039/c4ra13434f
102. Reddy KR, Nakata K, Ochiai T et al (2011) Facile fabrication and photocatalytic application of Ag nanoparticles-TiO_2 nanofiber composites. J Nanosci Nanotechnol 11:3692–3695. https://doi.org/10.1166/jnn.2011.3805
103. Reddy KR, Karthik KV, Prasad SBB et al (2016) Enhanced photocatalytic activity of nanostructured titanium dioxide/polyaniline hybrid photocatalysts. Polyhedron 120:169–174. https://doi.org/10.1016/j.poly.2016.08.029
104. Ren W, Ai Z, Jia F et al (2007) Low temperature preparation and visible light photocatalytic activity of mesoporous carbon-doped crystalline TiO_2. Appl Catal B Environ 69:138–144. https://doi.org/10.1016/j.apcatb.2006.06.015
105. Sahoo C, Gupta AK (2012) Optimization of photocatalytic degradation of methyl blue using silver ion doped titanium dioxide by combination of experimental design and response surface approach. J Hazard Mater 215–216:302–310. https://doi.org/10.1016/j.jhazmat.2012.02.072
106. Sahoo C, Gupta AK, Pillai IMS (2012) Heterogeneous photocatalysis of real textile wastewater: evaluation of reaction kinetics and characterization. J Environ Sci Health A Tox Hazard Subst Environ Eng 47:2109–2119. https://doi.org/10.1080/10934529.2012.695996
107. Sahoo C, Gupta AK, Sasidharan Pillai IM (2012) Photocatalytic degradation of methylene blue dye from aqueous solution using silver ion-doped TiO_2 and its application to the degradation of real textile wastewater. J Environ Sci Health A Tox Hazard Subst Environ Eng 47:1428–1438. https://doi.org/10.1080/10934529.2012.672387

108. Saleh TA, Gupta VK (2012) Photo-catalyzed degradation of hazardous dye methyl orange by use of a composite catalyst consisting of multi-walled carbon nanotubes and titanium dioxide. J Colloid Interface Sci 371:101–106. https://doi.org/10.1016/j.jcis.2011.12.038
109. Sampaio MJ, Marques RRN, Tavares PB et al (2013) Tailoring the properties of immobilized titanium dioxide/carbon nanotube composites for photocatalytic water treatment. J Environ Chem Eng 1:945–953. https://doi.org/10.1016/j.jece.2013.08.014
110. Sathish M, Viswanathan B, Viswanath RP, Gopinath CS (2005) Synthesis, characterization, electronic structure, and photocatalytic activity of nitrogen-doped TiO_2 nanocatalyst. Chem Mater 17:6349–6353. https://doi.org/10.1021/cm052047v
111. Scarisoreanu M, Fleaca C, Morjan I et al (2017) High photoactive TiO_2/SnO_2 nanocomposites prepared by laser pyrolysis. Appl Surf Sci 418:491–498. https://doi.org/10.1016/j.apsusc.2016.12.122
112. Shao J, Sheng W, Wang M et al (2017) In situ synthesis of carbon-doped TiO_2 single-crystal nanorods with a remarkably photocatalytic efficiency. Appl Catal B Environ 209:311–319. https://doi.org/10.1016/j.apcatb.2017.03.008
113. Sillanpää M, Ncibi MC, Matilainen A (2018) Advanced oxidation processes for the removal of natural organic matter from drinking water sources: a comprehensive review. J Environ Manage 208:56–76. https://doi.org/10.1016/j.jenvman.2017.12.009
114. Singh P, Vishnu MC, Sharma KK et al (2016) Photocatalytic degradation of Acid Red dye stuff in the presence of activated carbon-TiO_2 composite and its kinetic enumeration. J Water Process Eng 12:20–31. https://doi.org/10.1016/j.jwpe.2016.04.007
115. Singh B, Kaur G, Singh P et al (2017) Nanostructured BN–TiO_2 composite with ultra-high photocatalytic activity. New J Chem 41:11640–11646. https://doi.org/10.1039/C7NJ02509B
116. Sirirerkratana K, Kemacheevakul P, Chuangchote S (2019) Color removal from wastewater by photocatalytic process using titanium dioxide-coated glass, ceramic tile, and stainless steel sheets. J Clean Prod 215:123–130. https://doi.org/10.1016/j.jclepro.2019.01.037
117. Siuzdak K, Szkoda M, Lisowska-Oleksiak A et al (2015) Thin layer of ordered boron-doped TiO_2 nanotubes fabricated in a novel type of electrolyte and characterized by remarkably improved photoactivity. Appl Surf Sci 357:942–950. https://doi.org/10.1016/j.apsusc.2015.09.130
118. Song J, Wang X, Chen O-P et al (2015) Photocatalytic degradation of reactive black-5 dye with novel graphene-titanium nanotube composite. Sep Sci Technol 50:1394–1402. https://doi.org/10.1080/01496395.2014.969377
119. Stathatos E, Papoulis D, Aggelopoulos CA et al (2012) TiO_2/palygorskite composite nanocrystalline films prepared by surfactant templating route: synergistic effect to the photocatalytic degradation of an azo-dye in water. J Hazard Mater 211–212:68–76. https://doi.org/10.1016/j.jhazmat.2011.11.055
120. Štengl V, Bakardjieva S, Murafa N (2009) Preparation and photocatalytic activity of rare earth doped TiO_2 nanoparticles. Mater Chem Phys 114:217–226. https://doi.org/10.1016/j.matchemphys.2008.09.025
121. Sun J, Qiao L, Sun S, Wang G (2008) Photocatalytic degradation of Orange G on nitrogen-doped TiO_2 catalysts under visible light and sunlight irradiation. J Hazard Mater 155:312–319. https://doi.org/10.1016/j.jhazmat.2007.11.062
122. Suwarnkar MB, Dhabbe RS, Kadam AN, Garadkar KM (2014) Enhanced photocatalytic activity of Ag doped TiO_2 nanoparticles synthesized by a microwave assisted method. Ceram Int 40:5489–5496. https://doi.org/10.1016/j.ceramint.2013.10.137
123. Tahiri Alaoui O, Nguyen QT, Mbareck C, Rhlalou T (2009) Elaboration and study of poly(vinylidene fluoride)–anatase TiO_2 composite membranes in photocatalytic degradation of dyes. Appl Catal A Gen 358:13–20. https://doi.org/10.1016/j.apcata.2009.01.032
124. Tristão JC, Magalhães F, Corio P, Sansiviero MTC (2006) Electronic characterization and photocatalytic properties of CdS/TiO_2 semiconductor composite. J Photochem Photobiol A Chem 181:152–157. https://doi.org/10.1016/j.jphotochem.2005.11.018
125. Umebayashi T, Yamaki T, Tanaka S, Asai K (2003) Visible light-induced degradation of methylene blue on S-doped TiO_2. Chem Lett 32:330–331. https://doi.org/10.1246/cl.2003.330

126. UNESCO WWAP (2020) World water development report 2020—water and climate change
127. Upadhyay GK, Rajput JK, Pathak TK et al (2019) Synthesis of ZnO:TiO_2 nanocomposites for photocatalyst application in visible light. Vacuum 160:154–163. https://doi.org/10.1016/j.vacuum.2018.11.026
128. Varnagiris S, Medvids A, Lelis M et al (2019) Black carbon-doped TiO_2 films: synthesis, characterization and photocatalysis. J Photochem Photobiol A Chem 382: https://doi.org/10.1016/j.jphotochem.2019.111941
129. Wongso V, Chen CJ, Razzaq A et al (2019) Hybrid kaolin/TiO_2 composite: effect of urea addition towards an efficient photocatalyst for dye abatement under visible light irradiation. Appl Clay Sci 180: https://doi.org/10.1016/j.clay.2019.105158
130. Xiao Q, Zhang J, Xiao C et al (2008) Solar photocatalytic degradation of methylene blue in carbon-doped TiO_2 nanoparticles suspension. Sol Energy 82:706–713. https://doi.org/10.1016/j.solener.2008.02.006
131. Xu J, Ao Y, Fu D, Yuan C (2008) Low-temperature preparation of F-doped TiO_2 film and its photocatalytic activity under solar light. Appl Surf Sci 254:3033–3038. https://doi.org/10.1016/j.apsusc.2007.10.065
132. Xu S, Feng D, Shangguan W (2009) Preparations and photocatalytic properties of visible-light-active zinc ferrite-doped TiO_2 photocatalyst. J Phys Chem C 113:2463–2467. https://doi.org/10.1021/jp806704y
133. Xu J, Sun P, Zhang X et al (2014) Synthesis of N-doped TiO_2 with different nitrogen concentrations by mild hydrothermal method. Mater Manuf Process 29:1162–1167. https://doi.org/10.1080/10426914.2014.921697
134. Xu Q, Feng J, Li L et al (2015) Hollow $ZnFe_2O_4/TiO_2$ composites: high-performance and recyclable visible-light photocatalyst. J Alloys Compd 641:110–118. https://doi.org/10.1016/j.jallcom.2015.04.076
135. Yan N, Zhu Z, Zhang J et al (2012) Preparation and properties of ce-doped TiO_2 photocatalyst. Mater Res Bull 47:1869–1873. https://doi.org/10.1016/j.materresbull.2012.04.077
136. Yang K, Dai Y, Huang B (2007) Understanding photocatalytic activity of S- and P-doped TiO_2 under visible light from first-principles. J Phys Chem C 111:18985–18994. https://doi.org/10.1021/jp0756350
137. Yang G, Jiang Z, Shi H et al (2010) Study on the photocatalysis of F-S co-doped TiO_2 prepared using solvothermal method. Appl Catal B Environ 96:458–465. https://doi.org/10.1016/j.apcatb.2010.03.004
138. Yang X, Wang S, Sun H et al (2015) Preparation and photocatalytic performance of Cu-doped TiO_2 nanoparticles. Trans Nonferrous Met Soc China 25:504–509. https://doi.org/10.1016/S1003-6326%5b15%5d63631-7
139. Yew S-P, Tang H-Y, Sudesh K (2006) Photocatalytic activity and biodegradation of polyhydroxybutyrate films containing titanium dioxide. Polym Degrad Stab 91:1800–1807. https://doi.org/10.1016/j.polymdegradstab.2005.11.011
140. Yu JC, Yu J, Ho W et al (2002) Effects of F-doping on the photocatalytic activity and microstructures of nanocrystalline TiO_2 powders. Chem Mater 14:3808–3816. https://doi.org/10.1021/cm020027c
141. Zhang H, Lv X, Li Y et al (2010) P25-graphene composite as a high performance photocatalyst. ACS Nano 4:380–386. https://doi.org/10.1021/nn901221k
142. Zhang X, Chen W, Lin Z, Shen J (2011) Photocatalytic degradation of a methyl orange wastewater solution using titanium dioxide loaded on bacterial cellulose. Synth React Inorganic Met Nano-Metal Chem 41:1141–1147. https://doi.org/10.1080/15533174.2011.591359
143. Zhou W, Liu Q, Zhu Z, Zhang J (2010) Preparation and properties of vanadium-doped TiO_2 photocatalysts. J Phys D Appl Phys 43: https://doi.org/10.1088/0022-3727/43/3/035301
144. Zhou S, Xia L, Zhang K et al (2020) Titanium dioxide decorated natural cellulosic Juncus effusus fiber for highly efficient photodegradation towards dyes. Carbohydr Polym 232: https://doi.org/10.1016/j.carbpol.2020.115830
145. Zhu Y, Dan Y (2010) Photocatalytic activity of poly(3-hexylthiophene)/titanium dioxide composites for degrading methyl orange. Sol Energy Mater Sol Cells 94:1658–1664. https://doi.org/10.1016/j.solmat.2010.05.025

146. Zhu Y, Xu S, Yi D (2010) Photocatalytic degradation of methyl orange using polythiophene/titanium dioxide composites. React Funct Polym 70:282–287. https://doi.org/10.1016/j.reactfunctpolym.2010.01.007
147. Zhou S, Xia L, Zhang K (2020) Titanium dioxide decorated natural cellulosic Juncus effusus fiber for highly efficient photodegradation towards dyes. Carbohydr Polym 232:115830. https://doi.org/10.1016/j.carbpol.2020.115830

Nanocellulose-Based Membranes for the Removal of Dyes from Aquatic Systems

Nhamo Chaukura, Thato M. Masilompane, Mpho Motsamai, Abel Hunt, and Khanya V. Phungula

Abstract Conventional methods used in dye removal have limited efficiencies. Due to its eco-friendliness and other desirable characteristics, such as biodegradation, environmental friendliness, nontoxicity, excellent thermal and mechanical properties, and ease of modification, cellulose has found application in fabrication of membranes for dye removal from wastewater. Unlike the bulk form, the properties of nanocellulose include high mechanical strength and high surface area, giving it potential for the fabrication of high-efficiency membranes. This stems primarily from the abundant surface OH groups. Moreover, nanocellulose affords a high aspect ratio, a large population of active binding sites, resulting in high adsorption capacity for a range of pollutants including dyes. The major challenge in the design of nanocellulose-based membranes is ensuring adequate access to reactive sites, together with maintaining high flux and mechanical stability. Generally, incorporating nanomaterials into the membrane matrix reduces fouling. This chapter reviews literature on the use of nanocellulose-based membranes in the remediation of dye-polluted wastewater. The specific objectives are: (1) to evaluate literature on the synthesis and fabrication of nanocellulose-based membranes and (2) to link the physico-chemical properties of the membranes to their dye removal performance.

Keywords Biomaterials · Nanofiltration · Photodegradation · Wastewater treatment

1 Introduction

Dyes are recalcitrant to aerobic degradation, and stable to heat, light, and oxidants [1]. As a result, the treatment of dye-polluted wastewater is challenging, and consequently

N. Chaukura (✉) · M. Motsamai · A. Hunt · K. V. Phungula
Department of Physical and Earth Sciences, Sol Plaatje University, Kimberley, South Africa
e-mail: nhamo.chaukura@spu.ac.za

T. M. Masilompane
Molecular Sciences Institute, School of Chemistry, University of the Witwatersrand, Johannesburg, South Africa

© The Author(s), under exclusive license to Springer Nature Singapore Pte Ltd. 2021
S. S. Muthu and A. Khadir (eds.), *Novel Materials for Dye-containing Wastewater Treatment*, Sustainable Textiles: Production, Processing, Manufacturing & Chemistry,
https://doi.org/10.1007/978-981-16-2892-4_6

dyes persist in treated water. A number of treatment approaches have been applied in the remediation of dye-contaminated wastewater, and these include adsorption, coagulation, chemical oxidation, electrochemical, photocatalytic, and membrane separation processes [2]. Among these, membrane technology is an effective method for the remediation of dye-contaminated water [3]. Recently, there has been a shift towards green technologies with a reduced carbon footprint (refs). In this regard, membranes derived from biomaterials have gained significant research interest.

Biomaterials are abundant, cheap, and have desirable physico-chemical properties, thus they have potential as feedstock for synthesizing and fabricating nanocellulose membranes [4, 5]. They are composed of several organic compounds such as hemicellulose, lignin, and cellulose with polyphenolic moieties that influence the binding mechanisms of dyes [5]. Cellulose is the most abundant renewable biopolymer and can be obtained from a range of biowastes. Owing to its biodegradation, excellent mechanical and thermal properties, nontoxicity, and relative ease of modification, cellulose has received considerable attention in the design, synthesis, and fabrication of water treatment materials and devices [6].

The cellulose surface has oxygenated functional groups, and the carbon backbone can be easily modified and tailored to introduce properties that target certain pollutants [5, 7]. Moreover, it is easily modified into nanocellulose, which has a high aspect ratio that endows it with high removal capacities [3]. Thus, cellulose and its derivatives such as cellulose nanocrystals, nanofibers, nanofibrils, or nanowires have great promise in the removal of pollutants such as dyes from aqueous systems. Besides being used as powdered or granular adsorbents, nanocellulose can be fashioned into membranes using various methods. Previous studies have reported high dye removal efficiencies using nanocellulose-based membranes [2, 8, 9]. This chapter aims at reviewing literature on the use of nanocellulose-based membranes in dye remediation. Specifically, the objectives are: (1) to evaluate literature on the synthesis and fabrication of nanocellulose-based membranes and (2) to link the physico-chemical properties of the membranes to their dye removal performance. Knowledge gaps and opportunities for future research were identified.

2 Synthesis and Fabrication of Cellulose-Based Membranes

2.1 Precursor Materials

Cellulose (30–50%) is a carbohydrate that co-occurs with lignin (10–20%), hemicellulose (15–35%), and a small proportion of other plant components such as amino acids, essential oils, fat reducing agents, flavonoids, gum, saponins, starch, sterols, and sugars within the plant fibers [6]. With its strength and flexibility, lignin protects the plant from microbial invasion, while hemicellulose serves as an adhesive between the lignin and cellulose. In turn, cellulose confers mechanical stability and hydrophobic character to the plant cell wall [6]. It is readily available, environmental friendly, and nontoxic as well as inexpensive. Various cellulosic materials ranging from leaf fibers, bast, straw, grass to wood, agricultural wastes or by-products, oil palm, and other sources such as bacteria or animals have been studied [6, 10–15]. The chemical structure of cellulose provides sites for interaction with various pollutants and avails opportunities for functionalization using a range of reactive moieties. It consists of D-glucopyranose monomeric units with β-1, 4-glycosodic linkages, and is stabilized by hydrogen bonding [16, 17]. The abundant OH groups facilitate hydrogen bonding that confer a three-dimensional fibrous structure on the cellulose [6]. The hydrogen bonding stabilizes cellulose and confers axial stiffness to cellulose fibrils. In addition, the OH groups permit the modification of surface charge using biological, chemical, and mechanical approaches. Within the cellulose fibril structure, there are areas where cellulose chains are highly ordered and crystalline, and other regions that are of low order and amorphous [18]. Because of these remarkable properties and the potential to sequestrate pollutants from aqueous systems, cellulose has received increasing research interest.

The fabrication of nanomaterials from cellulose has been widely investigated. These nanomaterials are subsequently used to fabricate cellulose-based membranes. Common nanomaterials derived from cellulose include bacterial nanocellulose (BNC), cellulose nanocrystals (CNCs), and cellulose nanofibers (CNFs) which can be synthesized from a range of biomass precursors through acid hydrolysis, mechanical disintegration, and enzyme treatment, respectively [19]. In a previous study, CNCs had the largest surface area (138–226 m^2/g) relative to CNFs (146–219 m^2/g) and Chitosan-CNC composites (76–86 m^2/g) (Liu et al. 2014). The crystallinity of the materials increased due to the elimination of amorphous phases in cellulose sludge during the synthesis process. Thermal degradation experiments showed CNCs had the least degradation temperature (209 °C), followed by ChCNCs (200 °C) and CNFs (258 °C). The zeta potentials followed the trend CNC < CNF < Chitosan-CNC with values in the range of −44.4 to 47.3 mV, most of which were negative, suggesting negatively charged surfaces, save for the positively charged CNF surface at lower pH. Overall, the conversion of cellulose to nanocellulose produces materials with a high aspect ratio and modified surface chemistry, and thus suitable for subsequent modification.

2.1.1 Cellulose Nanofibrils

CNFs contain a network of amorphous and crystalline, micrometer long and nanometer wide, fibrils [19]. As a result, they possess a large specific surface area and high aspect ratio. Moreover, the high density of surface OH groups make CNFs suitable for modification [19]. Mechanical techniques like grinding, high-pressure homogenization, microfluidization, and ultrasonication are used to extract CNFs from lignocellulosic plants. Chemical or enzymatic pretreatments are commonly used to reduce the energy demand for making cellulose nanofibrils with smaller diameter and a narrower size distribution [8]. Such pretreatments include carboxymethylation, enzymatic hydrolysis, phosphorylation, and oxidation using 2,2,6,6-tetramethylpiperidine-1-oxyl (TEMPO). Of these, TEMPO oxidation is commonly used to isolate CNFs [4]. The properties of the resulting CNFs are influenced by the source of the cellulose material. In addition, the degree of crystallinity, the intensity of the fibrillation, morphology, and structural properties of the resulting CNFs are artefacts of the extraction method [4]. The large surface area and high mechanical strength increase the interaction with the secondary materials at nano-scale to produce a new functionality in hybrid materials [20].

2.1.2 Cellulose Nanocrystals

CNCs are the crystalline component of CNFs following the removal of the amorphous phase through acid hydrolysis. Their properties mainly depend on the hydrolysis conditions, which include the concentration of the acid, reaction temperature, and reaction time. Usually, CNCs have needle-like or rod-like geometries, with dimensions between 1 and 50 nm wide and few 100 nm long. They exhibit high surface areas, excellent thermal and mechanical properties, and a high high-magnetic response, and can potentially enhance the mechanical properties of hybrid materials [7, 14]. Moreover, they are hydrophilic owing to the abundant OH groups [21]. The CNCs possess a high negative surface charge density, and thus provide high adsorption capacity for the removal of positively charged pollutants [22]. Compared to CNFs, CNCs are rigid owing to the high degree of crystallinity.

As a result, CNCs have potential in enhancing membrane performance. For instance, previous studies have fabricated membranes from CNCs and polymers and metals, and reported improved and increased porosity, increased hydrophilicity, and in some cases higher selectivity, leading to higher water permeability [21]. In another study, the incorporation of CNCs into the interlayer of nanofiltration (NF) membranes resulted in high permeation and increased selectivity towards monovalent ions.

2.1.3 Bacterial Nanocellulose

Apart from CNCs and CNFs, BNCs are biosynthesized from a wide range of bacteria, including *Gluconacetobacter xylinus* and *Acetobacter xylinum*, via a bottom-up biofabrication assembly technique [16, 23]. Compared to other nanocellulose materials, BNCs possess the highest purity. The properties of BNCs include nanofiber network structures and a high moisture content (99%) in the form of thermally and mechanically stable hydrogels [16]. They possess twisted ribbon geometry with around 10–20 nm diameter and micrometer lengths [3]. The drying method significantly influences the properties. Air-drying results in the collapse of the three-dimensional microstructure as manifested through wrinkling due to fiber aggregation, shrinkage, and superficial hornification, resulting in reduced porosity [16]. Owing to its superior crystalline structure and higher purity relative to plant cellulose, BNCs have excellent mechanical properties [3]. Notably, BNC fibrils offer possibilities to obtain the nanofibrils naturally. Nevertheless, biomass derived from plants remain the most abundant and low-cost source of cellulose, especially in large-scale processes. Besides, plant-derived nanocellulose is more biodegradable, more hydrophilic, and thus more eco-friendly. In addition, the high aspect ratio lends it for adsorption applications [3, 16].

2.1.4 Pretreatment

The purpose of pretreatment is to remove impurities such as ash, hemicellulose, lignins, and wax to produce pure cellulose. Through chemical or mechanical treatment, cellulose can be converted to CNCs and CNFs, which are subsequently fabricated into membranes by a number of methods [23]. Mechanical processes such as grinding/crushing, high-pressure homogenization, and ultrasonication are some of the physical methods used in the preparation of CNFs. CNWs are synthesized via acid hydrolysis and are made up of short fibrils. However, this method is limited by low yield and shorter fiber lengths [23].

2.1.5 Chemical Modification of Cellulose

Chemical modification of cellulose can be achieved by either direct modification of OH groups through various modifying agents, or grafting monomers to cellulose. The existence of three OH groups in each monomer unit and their high reactivity give cellulose the properties such as hydrophilicity, chirality, and biodegradability [5, 17]. Besides, OH groups allow for surface modification and tuning of surface charge. Depending on the chemical treatment used, this produces cellulose derivatives that have tailored properties for targeted application [7]. Chemical modification of cellulose depends on the position attacked, which can either be anhydro-glucose units ring opening or attack on the OH group [4]. Apart from the OH groups, the CNC surface may contain other functional groups arising from the preparation and isolation conditions. Depending on the hydrolysis pathway, common functional groups introduced during the isolation stage include carboxyl and sulfate groups. These groups confer a negative charge on the CNC surface in the pH range exceeding the pK_a [1]. In contrast, CNFs are obtained through mechanical disruptions and are not functionalized and therefore, require further modification to widen their scope of application. The most frequently used hydrolyzing agent is H_2SO_4, which reacts with the surface OH groups of cellulose via an esterification reaction by grafting anionic sulfate esters. The introduction of these groups carrying a negative charge facilitates dispersion of the nanocellulose in aqueous solution through electrostatic repulsion [3]. By modifying the cellulose, physico-chemical properties and selective interaction improves. Using quaternary ammonium groups gives CNC/CNF a permanent cationic charge.

Grafting structural moieties onto cellulose using agents such as NaOH, imidazole, and amines can impart specific structural and physico-chemical properties to the modified cellulose [16]. Following membrane fabrication, functional groups can be grafted onto the surface of nanocellulose via amidation, carboxylation, esterification, etherification, phosphorylation, oxidation, and sulfonation [23]. Such membranes are selective in the retention of metal ions, dyes, and microbes in water treatment. Grafting polymers carrying reactive groups is a novel technique that produces membranes with high adsorption and rejection capacities for a wide range of pollutants. For instance, polymers carrying amino groups were grafted onto nanocellulose and showed a significant rise in the adsorption of a range of dyes owing to interactions between dye molecules and the amino group.

3 Membrane Fabrication

Nanocellulose is an excellent precursor for the fabrication of membranes owing to their excellent chemical and mechanical properties and their ability to form a range of materials with diverse structures and properties [9, 24]. Various types of nanocellulose have been used to fabricate membranes with tunable porosity [23]. Properties like selectivity and permeability are influenced by the chemistry and microstructure of the feedstock. A range of methods have been used to fabricate membranes from nanocellulose. These include solvent casting, spin-coating, and electrospinning, freeze-drying followed by compaction, vacuum-filtration followed by dip-coating [22, 25]. The method used determines the porosity and permeability of the resultant membrane, among other properties. The content of nanocellulose in the hybrid membrane has a bearing on the ensuing properties. For example, a previous study demonstrated that drying CNFs at higher CNC content constricts the nanocellulose network structure and reduces pore sizes [22].

Composite membranes are frequently used in ultrafiltration applications. Because thinner layers give a higher flux, ultrathin ultrafiltration membranes are the best high-flux separation membranes in terms of effectiveness. However, a balance has to be struck to maximize flux while optimizing the mechanical strength of the membrane.

3.1 Modification of Nanocellulose-Based Membranes

Cellulose can be chemically modified by either direct modification of OH groups through various modifying agents, or grafting monomers to cellulose. The existence of three OH groups in each monomer unit and their high reactivity give cellulose the properties such as hydrophilicity, chirality, and biodegradability [5, 17]. Besides, OH groups allow for surface modification and tuning of surface charge. Depending on the chemical treatment used, this produces cellulose derivatives that have tailored properties for targeted application [7]. Chemical modification of cellulose depends on the position attacked, which can either be anhydro-glucose units ring opening or attack on the OH group [4]. Apart from the OH groups, the CNC surface may contain other functional groups arising from the preparation and isolation conditions. Depending on the hydrolysis pathway, common functional groups introduced during the isolation stage include carboxyl and sulfate groups. These groups confer a negative charge on the CNC surface over a range of pH values exceeding the pK_a [1]. In contrast, CNFs are obtained through mechanical disruptions and are not functionalized and therefore require further modification to widen their scope of application. The most common hydrolyzing agent is H_2SO_4, which reacts with the cellulose OH groups through an esterification reaction by grafting anionic sulfate ester moieties. The introduction of these negatively charged moieties facilitates dispersion of nanocellulose in aqueous solution through electrostatic repulsion [3]. Thus, by modifying the cellulose, physico-chemical properties and selectivity improves.

4 Properties of Nanocellulose-Based Membranes

The properties of nanocellulose-based membranes are largely inherited from the precursor materials. These include tunable porosity and molecular weight cut-off, high hydrophobicity, and excellent mechanical properties (Table 1). In order to reduce the energy demand while keeping the high selectivity, a membrane should ideally have high flux at low pressure [9, 24]. Finding suitable membrane processing pathways to achieve these goals is however, a challenge. Owing to the bio-based nature, biodegradability, and functionality offered by cellulose nanomaterials, the development of nanocellulose-based membranes is an interesting area for both research and industry.

The mechanical properties of membranes are particularly important for the membrane to withstand the high operation pressures. Nanocellulose has been widely reported to enhance the mechanical strength and durability of hybrid membranes. For example, an investigation on CNCs-embedded natural rubber latex (NRL) composite reported a rise in the tensile strength (27.3%) and tensile modulus (24.8%) with increasing CNCs content (2–4%), whereas the elongation at break decreased (17.4%) [34]. The enhancement of mechanical properties was due to the OH groups in NRL chains which facilitated the interaction of the rubber matrix with CNCs. Another study developed a CNCs and chitosan composite membrane by freeze-drying, followed by compaction and subsequent cross-linking with glutaraldehyde vapors [25]. Both the pristine and cross-linked membranes had similar tensile strengths (1 MPa), while a 148.4% increase in the tensile modulus was observed after cross-linking. The membrane had an average pore diameter of 13.0 nm and reduced flux (64 L m^{-2} h^{-1}), and effectively retained cationic charged dyes, namely, Victoria Blue, Rhodamine, and Methyl Violet.

A previous study observed that dip-coating a CNF layer with CNCs increased the tensile strength and tensile modulus of the membrane [22]. Solvent treatment is a widely used strategy to tune porosity and surface area. For instance, acetone treatment increased the pores and reduced hydrogen bonding, and enhancing the membrane surface area as a result. Similarly, porosity can be tuned using nanomaterials such as

Table 1 Properties of cellulose-based membranes

Membrane synthesis	Properties	Comments	References
CNF template for magnetic nanoparticles synthesis via in situ hydrolysis	Size <20 mm, crystalline size 96–130 Å Paramagnetic properties: 67.4–38.5 emu g^{-1}	94.9% Rhodamide B degradation in 30 min	Amilian et al. (2020)
Filter paper coated with cellulose hydrogel layer by dip-coating	Water contact angle: ~0°. Smooth surface	Separates oil/water emulsion Addition of citric acid enhances mechanical properties and adsorption of methylene blue	[2]
CNFs loaded with reduced graphene oxide laminates via vacuum-filtration, membrane treated with oxygen plasma (1–4 min)	Pure water permeability: 37.2 ± 3.9 Lm^{-2} h^{-1} bar^{-1}	>90% for acid Fuschin, Rose Bengal, and Brilliant Blue	[26]
Nanocellulose membrane mercerized and modified with succinic anhydride	Carboxylic groups as active sites Crystalline	−0.72–1.95 mmol/g of metal uptake by order Cd > Cu > Zn > Co > Ni Separation mechanism: adsorption. Regeneration using HNO_3 and ultrasound	[27]
Multi-layered cellulose membranes fabricated via vacuum-filtration and dip-coating	CNFs had high tensile strength. Pore size: 74 Å (NF range) Acetone treatment increased pore size to UF range (194 Å) and 7000% decrease in surface area. Increased flux (0–25 $L.m^{-2}$ h^{-1})	Middle layer of CNF removed majority of resistant flux which can be improved by reducing layer thickness Removes Ag^{+}, Cu^{2+}, and Fe^{2+}/Fe^{3+}. Removal mechanisms: Surface adsorption and micro precipitation	[22]

(continued)

Table 1 (continued)

Membrane synthesis	Properties	Comments	References
Charged polyelectrolytes of sodium carboxymethyl cellulose on polypropylene (CMCNa/PP) thin-film composite membranes produced via dip-coating and crosslinking	Long term performance Anti-fouling property through submerged NF	Negatively charged CMCNa/PP removed anionic dye. Dye retention and water permeability influenced by trans-membrane pressure, dye, and salt concentration	[28]
Composite hollow fiber NF membranes fabricated through layer-layer deposition of oppositely charged CMCNa and polyethylenimine on polypropylene hollow fiber support, with glutaraldehyde cross-linking	Positively charged membrane with salt rejection: $MgCl_2$ > $CaCl_2$ > KCl > NaCl > $MgSO_4$ > Na_2SO_4 at pH 7 MWCO: 730 g/mol. Water permeability: 14.2 L/m^2 h.bar	Rejects $MgCl_2$ (93.3%) and NaCl (36.2%) in 500 mg/L salt solution at 3.0 bar. >90% retention of Brilliant green, Victoria blue B and Congo red. Applicable for high-flux NF	[29]
3-D BNC with integrated 2-D graphene oxide nanosheets to fabricate robust anti-fouling UF membrane modified with dopamine	Superhydrophilicity and underwater superoleophobicity Water flux: 1149.3 L m^{-2} h^{-1}. Contaminant resistant	Anti-fouling properties, flux recovery ratio: ~96.9%	[28]
Hollow fibers prepared by electrospinning using ionic liquids as solvents	Porous crystalline hydrated membranes	>90% rejection for Congo Red in C_2H_5OH, ~100% in H_2O Improved results achieved using 1-ethyl-3-methylimidazolium diethyl phosphate and 1,3-dimethylimidazolium phosphate	[30]

(continued)

Table 1 (continued)

Membrane synthesis	Properties	Comments	References
High-flux nanofibrous UF membrane with integrated nanocellulose in a polyacrylonitrile (PAN) barrier layer, and electrospun PAN nanofibers on polyethylene terephthalate (PET) non-woven mat for supporting layer	Good tensile strength and hydrophobic. Anti-fouling Water flux: 1504 L/m^2 h Maximum pore size: 20.8–46.0 mm	High permeation flux (147.1 L/m^2h), high rejection ratio (98.4%). High selectivity against protein and polypeptides. 2–3 times water permeation flux relative to commercial membranes	[31]
NCFs prepared from wood pulp via TEMPO oxidation. An UF thin-film nanofibrous composite membrane was fabricated from these CNFs as top layer, PAN electronspun scaffold as mid-layer, and PET non-woven as support	CNFs dependent on cellulose concentration, pH, and ionic strength of oxidized cellulose aqueous suspension. MWCO: 55 nm	Permeate flux fivefold superior to commercial UF membranes. For UF of oil/water emulsions, the permeate flux was eightfold greater than commercial PAN10 membrane	[19]
Ultrathin polymeric NF membrane fabricated through surface modification of ultrafine CNF	Hydrophilic. Thickness: 77.4 nm. Mean pore size: 0.45 nm. MWCO: 824 g/mol, 32.7 L/m^2 h bar	Positively charged membranes with high permeation flux, excellent rejection of salts and dyes	[32]
Membranes prepared via phase inversion from solutions of 1-ethyl-3-methylimidazolium acetate ([EMIM]OAc) as solvent and acetone as volatile co-solvent	20% cellulose-80% ([EMIM]OAc) had 0.3 $L/m^{2\cdot}$ h bar water flux. 94% Bromothymol Blue rejection 12% cellulose-63% ([EMIM]OAc) had 8.4 L/m^2 h bar water flux and 69.8% Bromothymol Blue rejection	The rejection of dyes of different charge and polarity was influenced by solute–membrane interactions. Removal was attributed to electrostatic and hydrogen bonding interactions	[33]

graphene oxide (GO). An ultrathin GO layer fabricated on a CNF membrane reduced the average pore size as the thickness of GO layer increased, indicating the GO layer had denser and smaller pores relative to the CNF layer. In addition, the membrane showed improved mechanical stability.

The porosity of a membrane determines the water permeability and flux. A study investigating the impact of CNC on the properties of polyethersulfone (PES)-based membranes used MWCO measurements to determine the membrane pore size (Lessan et al. 2020). Neat PES membranes exhibited higher rejection, while the membranes loaded with CNC (5 wt%) had poor performance, showing the least rejection on all PEG molecular weights. Higher rejection rates for the 1 and 1.5 wt% PES CNC were observed from molecular weights between 4 and 35 KDa in an ascending order. At 2 wt% CNC content, the rejection rate increased linearly from 1 to 35 KDa.

5 Cellulose-Based Membranes for Dye Removal

Dyes are recalcitrant to aerobic digestion, biodegradation, and light. Although biological, chemical, and physical approaches are commonly used for dye removal from wastewater, membrane filtration is the most effective. The performance of nanocellulose-based membranes is dependent on the pore structure and the flow mode used. The performance of CNFs membranes is mainly associated with their selectivity in adsorbing pollutants from water, which is determined by surface functional groups. The intrinsic hydrophobicity can reduce fouling, which, along with high cost and limited flux, are major challenges in membrane technology [23]. Thus, efficient technologies for clean water recovery from dye-polluted wastewater are highly significant. Nanofiltration (NF), a pressure-driven membrane technique with characteristics between those of ultrafiltration (UF) and reverse osmosis (RO) membranes, is the most promising option for removing dyes from wastewater [35].

A number of previous studies have reported the use of nanocellulose-based membranes in dye removal from polluted wastewater (Table 2). For example, sulfated CNCs in a chitosan matrix was reported to remove 98, 84, 69% of Victoria Blue 2B, Methyl Violet 2B, and Rhodamine 6G, respectively [25]. In another study, zwitterionic BNC crosslinked with poly(2-methacryloyloxyethyl phophorylchline) performed 5.7–9.1 times better than pristine BNC in removing methyl orange and methylene blue dyes [36, 37]. However, for nanocellulose hybrid crosslinked with polyhedral oligomeric silsequioxane containing multi-N-methylol, the adsorption for Reactive yellow B-4RN and Reactive Blue BRN dyes did not exceed 10% [38]. Adsorption increased with an increase in contact time. A study using carboxylic functionalized CNCs achieved a maximum Crystal violet dye adsorption of 244 mg/g [39]. While the CNC membrane produced by oxidation of sodium periodate reacted with ethylenediamine had an Acid red GR dye uptake of 556 mg/g [40, 41], crosslinked microgel polyvinylamine CNC had removal capacities of 896, 1469, and 1250 mg/g for Acid red, Congo red, and Reactive light yellow, respectively [42].

Table 2 Applications of cellulose-based membranes in dye removal and comparison with other membranes/methods

Dye	Membrane	% Removal	Comments	References
Victoria Blue 2B, Methyl Violet 2B, Rhodamine 6G	Sulfated CNCs in chitosan matrix through freeze-drying followed by compacting, then cross-linked with gluteraldehyde vapor	98, 84, 69%	Electrostatic attraction between negatively charged CNCs and cationic dyes	[25]
Methylene blue, methyl orange	Zwitterionic BNC cross-linked with poly(2-methacryloyloxyethyl phophorylchline) (PMPC)	PMPC/BNC perform 5.7–9.1 times better than pure BNC in methylene blue and methyl orange removal. Amino-functionalized aerogels had better performance for methyl orange removal (266 mg/g)	Electrostatic interaction between both cationic and anionic dyes and the reduction of both gram-positive and gram-negative pathogenic bacteria	[36, 37]
Reactive yellow B, Reactive Blue	Nanocellulose hybrid cross-linked with polyhedral oligomeric silsequioxane containing multi-*N*-methylol	Adsorption for both dyes $\leq$10%. Adsorption increased with contact time	Adsorption was between dyes and the -C-N- function groups and the cubic cores in the nano-cellulose composites	[38]
Acid red	Oxidation of sodium periodate sequentially reacted with ethylenediamine to make CNC	Dye uptake: 556 mg/g	Maximum dye uptake was H-dependent	[40, 41]
Acid red Congo red Reactive light yellow	Crosslinked microgel polyvinylamine CNC	Maximum uptake: 896, 1469, and 1250 mg/g for Acid red, Congo red, and Reactive light yellow, respectively		[42]

(continued)

Table 2 (continued)

Dye	Membrane	% Removal	Comments	References
Crystal violet	CNC with carboxylic groups introduced via esterificaton of surface OH groups using maleic anhydride	Maximum adsorption: 244 mg/g	Electrostatic interactions between negatively charged CNC membrane and cationic dyes	[39]

Besides molecular exclusion due to intrinsic porosity, nanocellulose-based membranes also remove pollutants through adsorption, the mechanism of which mainly depends on the nature and availability of active groups serving as adsorption sites, whereas the membrane pore structure determines the accessibility to the active sites.

6 Conclusion and Future Outlook

In contrast to the bulk form, nanocellulose has high mechanical strength and a high surface area, making it very promising for producing membranes. Owing to their ease of modification and excellent dye retention properties, cellulose-based membranes have been extensively applied for the remediation of dye-polluted wastewater. Besides molecular exclusion due to intrinsic porosity, nanocellulose-based membranes also remove pollutants through adsorption, which depends on the nature and availability of adsorption sites.

Generally, membranes are susceptible to fouling owing to hydrophobicity, causing significant decrease in pollutant retention efficacy, flux, and lifespan. While incorporation of nanomaterials into the membrane matrix has been reported to reduce fouling, more strategies need to be developed to prolong the lifespan of the membrane.

Further studies should thus focus on:

(1) developing sustainable synthesis and fabrication methods to produce low-cost membranes. So far this has been pursued through the use of biowastes as feedstock for nanocellulose.
(2) methods for tailoring membrane porosity without compromising flux and consequently pollutant removal.
(3) reduction of fouling through cost-effective intelligent processing techniques.

References

1. Soon CY, Rahman NA, Tee YB, Talib RA, Tan CH, Abdan K, Chan EWC (2019) Electrospun biocomposite: nanocellulose and chitosan entrapped within a poly(hydroxyalkanoate) matrix for Congo red removal. J Mater Res Technol 8(6):5091–5102
2. Ao C, Zhao J, Li Q, Zhang J, Huang B, Wang Q, Gai J, Chen Z, Zhang W, Lu C (2020) Biodegradable all-cellulose composite membranes for simultaneous oil/water separation and dye removal from water. Carbohydr Polym 250:116872
3. Tan H-F, Ooi BS, Leo CP (2020) Future perspectives of nanocellulose-based membrane for water treatment. J Water Process Eng 37:101502
4. Farooq, A, Patoary MK, Zhang M, Mussana H, Li M, Naeem MA, Mushtaq M, Farooq A, Liu L (2020) Cellulose from sources to nanocellulose and an overview of synthesis and properties of nanocellulose/zinc oxide nanocomposite materials. Int J Biol Macromol 154:1050–1073
5. Thomas P, Duolikun T, Rumjit NP, Moosavi S, Lai CW, Johan MRB, Fen LB (2020) Comprehensive review on nanocellulose: Recent developments, challenges and future prospects. J Mech Behav Biomed Mater 110:103884
6. Dai H, Ou S, Huang Y, Huang H (2018) Utilization of pineapple peel for production of nanocellulose and film application. Cellulose 25:1743–1756
7. Tshikovhi A, Mishra SB, Mishra AK (2020) Nanocellulose-based composites for the removal of contaminants from wastewater. Int J Biol Macromol 152:616–632
8. Amiralian N, Mustapi M, Hossain MSA, Wang C, Konarova M, Tang J, Na J, Khan A, Rowan A (2020) Magnetic nanocellulose: A potential material for removal of dye from water. J Hazard Mater 394:122571
9. Yin X, Zhang Z, Ma H, Venkateswaran S, Hsiao BS (2020) Ultra-fine electrospun nanofibrous membranes for multicomponent wastewater treatment: Filtration and adsorption. Sep Purif Technol 242:116794
10. Ditzel FI, Prestes E, Carvalho BM, Demiate IM, Pinheiro LA (2017) Nanocrystalline cellulose extracted from pine wood and corncob. Carbohyd Polym 157:1577–1585
11. Feng X, Meng X, Zhao J, Miao M, Shi L, Zhang S, Fang J (2015) Extraction and preparation of cellulose nanocrystals from dealginate kelp residue: structures and morphological characterization. Cellulose 22:1763–1772
12. Johar N, Ahmad I, Dufresne A (2012) Extraction, preparation and characterization of cellulose fibres and nanocrystals from rice husk. Ind Crop Prod 37:93–99
13. Khawas P, Deka SC (2016) Isolation and characterization of cellulose nanofibers from culinary banana peel using high intensity ultrasonication combined with chemical treatment. Carbohydr Polym 137:608–616
14. Neto WPF, Silvério HA, Dantas NO, Pasquini D (2013) Extraction and characterization of cellulose nanocrystals from agro-industrial residue-Soy hulls. Ind Crop Prod 42:480–488
15. Santos RMD, Neto WPF, Silvério HA, Martins DF, Dantas NO, Pasquini D (2013) Cellulose nanocrystals from pineapple leaf, a new approach for the reuse of this agrowaste. Ind Crop Prod 50:707–714
16. Klemm D, Cranston E, Fischer D, Gama M, Kedzior S, Kralisch D, Kramer F, Kondo T, Lindström T, Nietzsche S, Petzold-Welcke K, Rauchfuß F (2018) Nanocellulose as a natural source for groundbreaking applications in materials science: today's state. Mater Today 21:720–748
17. Shahnaz T, Fazil MM, Padmanaban VC, Narayanasamy S (2020) Surface modification of nanocellulose using polypyrrole for the adsorptive removal of Congo red dye and chromium in binary mixture. Int J Biol Macromol 151:322–332
18. Poletto M, Pistor V, Zattera AJ (2013) Structural characteristics and thermal properties of native cellulose. In: Cellulose—fundamental aspects, Chapter 2. IntechOpen, pp 45–68
19. Ma H, Burger C, Hsiao BS, Chu B (2014) Fabrication and characterization of cellulose nanofiber based thin-film nanofibrous composite membranes. J Membr Sci 454:272–282

20. Khalil HPS, Saurabh CK, Adnan AS, Fazita MR, Syakira MI, Davoudpoura Y, Rafatullah M, Abdullah CK, Haafiz MKM, Dungani R (2016) A review on chitosancellulose blends and nanocellulose reinforced chitosan biocomposites: Properties and their applications. Carbohydr Polym 150:216–226
21. Bai L, Liu Y, Bossa N, Ding A, Ren N, Li G, Liang H, Wiesner MR (2018) incorporation of cellulose nanocrystals (CNCs) into the polyamide layer of thin-film composite (TFC) nanofiltration membranes for enhanced separation performance and antifouling properties. Environ Sci Technol 52:11178–11187
22. Karim Z, Claudpierre S, Grahn M, Oksman K, Mathew AP (2016) Nanocellulose based functional membranes for water cleaning: Tailoring of mechanical properties, porosity and metal ion capture. J Membr Sci 514:418–428
23. Wang Z, Zhang W, Yu J, Zhang L, Liu L, Zhou X, Huang C, Fan Y (2019) Preparation of nanocellulose/filter paper (NC/FP) composite membranes for high-performance filtration. Cellulose 26:1183–1194
24. Ang MBMY, Devanadera KPO, Duena ANR, Luo Z, Chiao Y, Millare JC, Aquino RR, Huang S, Lee K (2021) Modifying cellulose acetate mixed-matrix membranes for improved oil–water separation: Comparison between sodium and organo-montmorillonite as particle additives. Membranes 11:80. https://doi.org/10.3390/membranes11020080
25. Karim Z, Matthew AP, Grahn M, Mouzan J, Oksman K (2014) Nanoporous membranes with cellulose nanocrystals as functional entity in chitosan: Removal of dyes from water. Carbohydr Polym 112:668–676
26. Mohammed S, Hegab H, Ou R, Liu S, Ma H, Chen X, Sridhar T, Wang H (2020) Effect of oxygen plasma treatment on the nanofiltration performance of reduced graphene oxide/cellulose nanofiber composite membranes. Green Chem Eng. https://doi.org/10.1016/j.gce.2020.12.001
27. Hokkanen S, Repo E, Sillanpää M (2013) Removal of heavy metals from aqueous solutions by succinic anhydride modified mercerized nanocellulose. Chem Eng J 223:40–47
28. Yu S, Chen Z, Cheng Q, Lü Z, Liu M, Gao C (2012) Application of thin-film composite hollow fiber membrane to submerged nanofiltration of anionic dye aqueous solutions. Sep Purif Technol 88:121–129
29. Chen Q, Yu P, Huang W, Yu S, Liu M, Gao C (2015) High-flux composite hollow fiber nanofiltration membranes fabricated through layer-by-layer deposition of oppositely charged crosslinked polyelectrolytes for dye removal. J Membr Sci 492:312–321
30. Falca G, Musteata V, Behzad AR, Chisca S, Nunes SP (2019) Cellulose hollow fibers for organic resistant nanofiltration. J Membr Sci 586:151–161
31. Liang Y, Ma H, Taha AA, Hsiao BS (2020) High-flux anti-fouling nanofibrous composite ultrafiltration membranes containing negatively charged water channels. J Membr Sci 612:118382
32. Soyekwo F, Zhang Q, Gao R, Yan Qu, Lin C, Huang X, Zhu A, Liu Q (2017) Cellulose nanofiber intermediary to fabricate highly-permeable ultrathin nanofiltration membranes for fast water purification. J Membr Sci 524:174–185
33. Sukma FM, Çulfaz-Emecen PZ (2018) Cellulose membranes for organic solvent nanofiltration. J Membr Sci 545:329–336
34. Nair SS, Zhu JY, DengY RAJ (2014) High performance green barriers based on nanocellulose. Sustain Chem Process 2:23
35. Bai L, Liu Y, Ding A, Ren N, Li G, Liang H (2019) Surface coating of UF membranes to improve antifouling properties: a comparison study between cellulose nanocrystals (CNCs) and cellulose nanofibrils (CNFs). Chemosphere 217:76–84
36. Tang J, Song Y, Zhao F, Spinney S, Bernardes JS, Tam KC (2019) Compressible cellulose nanpfibril (CNF) based aerogels produced via a bio-inspired strategy for heavy metal ion and dye removal. Carbohydr Polym 208:404–412
37. Vilela C, Moreirinha C, Almeida A, Silverstre AJD, Freire CSR (2019) Zwitterionic nanocellulose-based membranes for organic dye removal. Materials 12(9):1404–1420
38. Xie K, Zhao W, He X (2011) Adsorption properties of nano-cellulose hybrid containing polyhedral oligomeric silsesquioxane and removal of reactive dyes from aqueous solution. Carbohydr Polym 83:1516–1520

39. Qiao H, Zhou Y, Yu F, Wang E, Min Y, Huang Q, Pang L, Ma T (2016) Effective removal of cationic dyes using carboxylated-functionalized cellulose nanocrystals. Chemosphere 141:297–303
40. Jin L, Li W, Xu Q, Sun Q (2015) Amino-functionalized nanocrystalline cellulose as an adsorbent for anionic dyes. Cellulose 22:2443–2456
41. Jin L, Sun Q, Xu Q, Xu Y (2015) Adsorptive removal of anionic dyes from aqueous solutions using microgel based on nanocellulose and polyvinylamine. Bioresour Technol 197:348–355
42. Zhu W, Liu L, Liao Q, Chen X, Qian Z, Shen J, Liang J, Yao J (2016) Functionalization of cellulose with hyperbranched polyethylenimine for selective dye adsorption and separation. Cellulose 23:3785–3797
43. Muqeet M, Mahar RB, Gadhi TA, Halima NB (2020) Insight into cellulose-based-nanomaterials—a pursuit of environmental remedies. Int J Biol Macromol 163:1480–1486
44. Putro JN, Santoso SP, Soetaredjo FE, Ismadji S, Ju Y (2019) Nanocrystalline cellulose from waste paper: adsorbent for azo dyes removal. Environ Nanotechnol Monit Manag 12:100260.
45. Teo HL, Wahab RA (2020) Towards an eco-friendly deconstruction of agro-industrial biomass and preparation of renewable cellulose nanomaterials: a review. Int J Biol Macromol 161:1414–1430
46. Wang DA (2019) A critical review of cellulose-based nanomaterials for water purification in industrial processes. Cellulose 26:687–701
47. Hu Y, Yue M, Yuan F, Yang L, Chen C, Sun D (2021) Bio-inspired fabrication of highly permeable and anti-fouling ultrafiltration membranes based on bacterial cellulose for efficient removal of soluble dyes and insoluble oils. J Membr Sci 621:118982

Current Treatment of Textile Dyes Using Potential Adsorbents: Mechanism and Comparative Approaches

A. Bennani Karim, H. Tounsadi, Y. Gaga, M. Taleb, Z. Rais, and N. Barka

Abstract Industrial effluents of dyes is a vital source of water pollution, their release from textile industries into surface water affects the ecosystem by generating oversized volumes of outlets mixed with several dyes. These dyes cause harmful health effects and lead to significant health concerns to humans and affect the environment. This chapter implements a contribution to environment in a sustainable view through adsorption and biosorption approaches to remove the methylene blue (MB) dye from aqueous solutions and from a real effluent of textile industry. In this survey, efficiencies of several adsorbents such as mineral, organic, synthetic, and low-cost materials are established. A brief insight into methylene blue dye removal mechanism and comparison among various adsorbents—Duste Apatite (DA), Phosphogypsium (PG), Raw Clay (RC), *Glebionis coronaria L.* (*G. coronaria L.*), *Diplotaxis Harra* (*D. Harra*), acidic and basic sawdust acacia (A-HCl and A-NaOH)—along with their sorption properties and their characteristics are discussed. Due to its good sorption capacity, chemically treated acacia tree sawdust has been successfully applied for eliminating textile dyes from wastewater. A real final effluent of a textile industry was thus treated by sorption on both acidic and basic chemically treated acacia tree sawdust.

Keywords Adsorption · Biosorption · Toxic dyes · Low-cost materials · Kinetic sorption · Sorption capacities

---

A. B. Karim · H. Tounsadi (✉) · M. Taleb · Z. Rais
Laboratoire d'Ingénierie des Matériaux Organométalliques, Moléculaires et Environnement (LIMOME), Faculté des Sciences Dhar EL Mahraz, Université Sidi Mohamed Ben Abdallah, Fès, Maroc
e-mail: hanane.tounsadi@usmba.ac.ma

Y. Gaga
Laboratory of Biotechnology, Conservation, and Valorization of Natural Resources -LBCVRN, Faculty of Sciences Dhar Mahraz, Sidi Mohamed Ben Abdellah University, Fez, Morocco

H. Tounsadi · N. Barka
Sultan Moulay, Research Group in Environmental Sciences and Applied Materials (SEMA), Slimane University of Beni Mellal, FP Khouribga, B.P. 145, Khouribga, 25000 Settat, Morocco

© The Author(s), under exclusive license to Springer Nature Singapore Pte Ltd. 2021
S. S. Muthu and A. Khadir (eds.), *Novel Materials for Dye-containing Wastewater Treatment*, Sustainable Textiles: Production, Processing, Manufacturing & Chemistry,
https://doi.org/10.1007/978-981-16-2892-4_7

1 Introduction

Toxic organic dyes are considered the major pollutants in wastewater discharged from textile and other industries. Due to their accumulation in the environment, these toxic substances cause various harm to the ecosystems. In effect, the release of huge colored wastewater into the ecosystem involves environmental damages such as esthetic pollution and troubles of aquatic life. Among 7×10^5 tons and approximately 10,000 different types of dyes and pigments produced worldwide annually, it is estimated that 1–15% of the dye is lost in the effluents during the dyeing process [61].

Furthermore, dyeing 1 kg of cotton could generate 200 kg of wastewater containing up to 50% of the initial input of dyes in the dye bath and up to 100 g/L of salts [41, 50]. The majority of these dyes are toxic, mutagenic, carcinogenic, and cause a serious water pollution.

Due to its effectiveness in various industrial fields, methylene blue (MB) can contribute to eye burns. Its inhalation can give rise to breathing difficulties and its ingestion through the mouth produces a burning sensation, causing nausea, vomiting, sweating, and profuse cold sweating [8]. Its treatment is proven to be effective and of great attention.

Therefore, to remove dyes from colored aqueous solutions, various processes have been used such as ultrafiltration [29], photocatalytic degradation [5], photocatalysis [4, 7], electrochemical treatments [23], coagulation/filtration [60], ozonation [21], reverse osmosis [2, 46], advanced oxidation processes [20, 34, 35], filtration [53], chemical precipitation [19, 52], electrocoagulation/flotation [42], biologic treatments [32], aerobic and anaerobic microbial degradation [28], ion exchange [26], and adsorption process [14, 34].

Howover, the majority of these techniques have exhibited their effectiveness to remove dyes, while, their industrial applications are limited because they are costly, require high energy, and time consumption. Then these methods can release large quantities of secondary sludge, which should be also properly treated in order to prevent the environmental pollution [27]. Wheras, adsorption remains a relatively easy technique to implement, it is widely used for water treatment [10, 47, 44, 54]. It has long been admited as a simple and economical technique for the treatment of textile dyes.

Due to its uncomplicated operation process, adsorption onto activated carbon has been considered the most efficient and applicable approach [36]. While, the utilization of activated carbon has some drawbacks such as the high cost and regeneration difficulties [3].

This work aims at the adsorption process of Methylene blue (MB) onto available low-cost adsorbents including Apatitic phosphate studied for the elimination of certain hardly degradable organic compounds such as metal ions and dyes [15, 17]; phosphogypsum, attempted during the last decade in agriculture, wastewater treatment, manufacture of plaster, cement, road embankments and earthworks, which have also been reused for the production of sulfuric acid [37], natural clays, in particular illite and kaolinite, reveald a great result in the treatment of colored effluents [9, 25, 40, 44]. *Diplotaxis harra (D. harra)* and *Glebionis coronaria L. (G. coronaria L.)* biomasses as plentiful, easily available, and no-toxic biosorbents and acidic and basic sawdust of acacia.

Biosorption and adsorption processes were investigated through the optimization of the contact time and adsorbent dosage. Equilibrium adsorption were modeled by Langmuir and Freundlich isotherm models. The surface properties of adsorbents were evaluated by FTIR, BET, SEM, and XRD methods.

Owing to its good sorption properties, higher chemical stability, and greater surface area, chemically treated sawdust of acacia tree has been successfully applied to remove textile dyes from wastewater.

2 Materials and Methods

2.1 Materials

The chemical used in this study was of analytical grade. MB (85%) was purchased from Panreac (Spain). The chemical structure and physical properties of this dye are shown in Table 1. NaOH was purchased from Merck (Germany), HNO_3 from Scharlau (Spain), and NaCl (99.5%) from Sigma–Aldrich (Germany).

2.2 Preparation of Adsorbent Materials

D. harra and *G. coronaria L.* were collected from the region of Khouribga in Morocco. They were sun-dried for 3 weeks, cut into small portions, and then were powdered to particle sizes of 120 μm using a domestic mixer. The residual powder of plants were dried in an oven at 70 °C for 24 h and stored for further utilization.

The acacia tree was purchased from the Azrou region in Morocco. The cut branches part of the acacia tree are used in adsorption. They are washed with the

Table 1 Chemical structure and physical properties of methylene blue

Dye	Chemical structure	Molar mass (g/mole)	λ_{max} (nm)	Area of use
MB	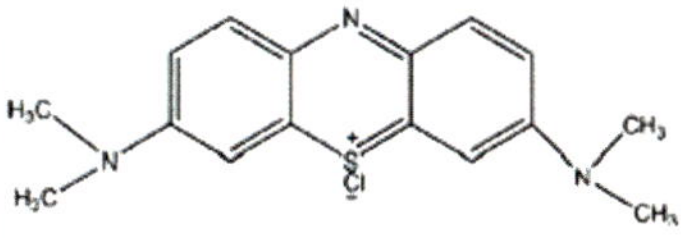	319.85	666	Textile industries, skin tanning, dyeing, food, or in the analysis of construction materials or as a colored indicator

water to remove adherent impurities, then, they are crushed and sieved to small particles with size ranging from 50 to 500 μm. The final powder has a sweet corn color. A quantity of 20 g of the acacia powder was chemically treated with 200 mL of NaOH (1 M) or HCl (1 M) for 3 h. Both obtained products are washed several times to remove traces of NaOH and HCl. Moreover, they are dried in an oven at 80 °C for 24 h. Lastly, two biosorbents based on acacia tree are acquired as A-NaOH and A-HCl.

The raw clay (RC) used in this work is supplied from a natural basin in the region of Safi (Morocco), crushed, and sieved to 0.08–0.1 μm size fractions. Further, it was dried at 378 K (105 °C) for 24 h and used for next applications.

The other adsorbents used in this work are the dust from natural apatites (DA) and the phosphogypsum (PG), ground, sieved to a particle size between 25 and 315 μm, and then dried for 24 h. The drying temperature was 105 °C for DA and 40 °C for PG.

2.3 *Characterization of Adsorbent Materials*

FTIR transmittance spectra of adsorbents were analyzed in the region of 4000–400 cm^{-1} using an ATR Miracle Diamante spectrometer by contacting a crystal (diamond) with the test sample without dilution in Ker matrix.

Specific surface area was evaluated by N2 adsorption according to the BET method using a Micromeritics model TRISTAR 3020 Instrument.

The morphological characteristics of adsorbents were investigated by scanning electron microscopy using a FEI Quanta 200 model. Small amount of each sample was finally powdered and mounted directly onto an aluminum sample holder using two-sided adhesive carbon model.

To identify the predominant phases of the clay mineral, X-ray diffraction analyses were performed using the Siemens D-5000 Diffractometer. The patterns were reported in 2θ range from 10 to 70° using a Bruker-axs D2-phaser advance diffractometer operating at 30 kV and 10 mA with CuKα. The interlayer spacing d002 was determined using the Bragg equation, $d = \lambda/2 \sin\theta$, where, λ is the X-ray length and θ is the scattering angle for the peak position.

2.4 *Sorption Studies*

2.4.1 Methylene Blue (MB) Sorption

Desired solutions were prepared by dissolving required weight of MB dye in distilled water and necessary concentrations were obtained by dilution. *D. harra* and *G. coronaria L.*, A-NaOH and A-HCl, RC, PA, and PG adsorption experiments were carried out in a series of 50 mL (for biosorption experiments) and 100 mL (for adsorption experiments) beakers respectively, containing the intended weight of each biosorbent and 50 mL of the dye solution at desired concentration. Adsorbent dosage from 0.5 to 10 g/L for *D. harra* and *G. coronaria L.* adsorption and A-NaOH and A-HCl, 0.1 g/L to 0.6 g/L for the RC, and 1 to 7 g/L for the PG and PA were added to the MB dye solution. Initial dye concentrations ranges from 20 to 200 mg/L, 10 to 500 mg/g for *D. harra* and *G. coronaria L.*, A-NaOH and A-HCl respectively, 10 to 60 mg/L for RC, PA, and PG and temperature from 10 to 50 °C.

Following each sorption experiment, the solid phase was separated from the liquid phase by centrifugation at 3,000 rpm for 10 min for *D. harra* and *G. coronaria L.*, RC, PA, and PG sorption and by filtration for 10 min for A-NaOH and A-HCl biosorption. Each sample was diluted by distilled water and the residual concentrations were determined from UV–vis characteristics at maximum absorption wavelength of MB dye (Table 1) using a TOMOS V-1100 UV–vis spectrophotometer. The absorbance of the treated MB stained solution was measured using a spectrophotometer [GBC (Ajax, Ontario) UV/visible 911] at the wavelength corresponding to the maximum dye absorption $\lambda_{max} = 666$ nm.

The adsorption capacity and adsorption yield were calculated using the following Eqs. (1) and (2):

$$q_t = \frac{[(C_0 - C_t) \times V]}{m} \quad (1)$$

where qt is the adsorbed quantity (mg/g), C_0 is the initial dye concentration (mg/L), C is the dye concentration at a time t (mg/L), and m is the mass adsorbents per liter of solution (g/L).

$$\%\ \text{removal} = \frac{(C_0 - C)}{C_0} \quad (2)$$

2.4.2 Industrial Effluent Biosorption

Final discharge of an industrial effluent from Sidi Brahim industry in Fez city was considered the "sampling site. The sampling was performed by the composite method to know the variations of the pollution characteristics during a day.

Three samples were implemented, and each sample is a mixture of six separate samples for each hour, of equal volumes. In the last of the day, à calculation makes it possible to constitute a composite sample by the association of 6 samples. Furthermore, three samples were achieved for the results reproductibility.

The water samples were collected with sterile vials during the spring season of 2018. Then, they were stored in a cooler at 4 °C during transport to the laboratory and then analyzed after 24 h. Then, water temperature, electrical conductivity, and pH were measured in situ. The water samples and their methods of analysis are those recommended by AFNOR and enacted by Rodier [49]. Samples were evaluated through various physicochemical analyses including pH, conductivity, temperature, chemical oxygen demand (COD), biological oxygen demand (BOD_5), NTK, suspended matter (MES), and heavy metals (ML) (Cd, Cr, Cu, Ni, and Zn). The heavy metals were evaluated by inductively coupled plasma spectrometry (ICP AES) (Ultima 2_jobin yvon).

3 Results and Discussion

3.1 Characterization of Adsorbent Materials

3.1.1 Infrared Spectroscopy

Infrared spectroscopy analysis of the chemically treated sawdust acacia, raw acacia, RC, AD, PG, and *D. harra* and *G. coronaria L.* are established in Fig. 1. The spectrum of biosorbents based on acacia shows a band around 3336.54 cm^{-1} indicating the presence of OH hydroxyl groups of cellulose, lignins, and hemicellulose components of wood. The bands at 2902.5 cm^{-1} correspond to C–H functions. The bands at 1266.45 are assigned to C–O functions. The bands at 1023.19 cm^{-1} can be corresponding to C–O–C functions. The bands appearing at the frequency between 720 and 400 cm^{-1} are characteristic of the C–H group in cellulose (Alemdar and M. Sain,

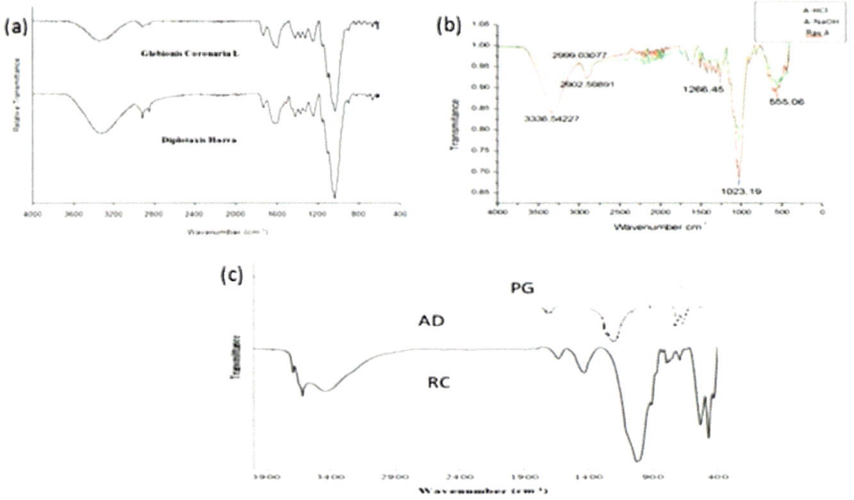

Fig. 1 Infrared absorption spectra of *G. coronaria L.* and *D. Harra* Biosorbents (**a**); Raw-A, A-HCl, and A-NaOH biosorbents (**b**); Red Clay, Apatites Dust, Phosphogypsum (**c**)

2008). From the spectrum, it could be deduced that the treated and untreated sawdust acacia tree contribute approximately the same bands with different intensities.

The infrared spectra of *D. harra* and *G. coronaria L.* shows broad absorption band at 3000–3600 cm^{-1} with a maximum at about 3333.8 cm^{-1} for *D. harra* and 3334.7 cm^{-1} for *G. coronaria L.* This band is characteristic of the stretching vibration of hydrogen bending of the hydroxyl group linked in cellulose, lignin, adsorbed water, and N–H Stretching [38]. The bands at 2917 and 2849.2 arise from aliphatic C–H stretching in an aromatic methoxyl group, in methyl, and methylene groups of side chains. The ν(OH−), ν(N–H), and ν(C–H) stretching is larger in the case of *D. harra* than of *G. coronaria L.* The small band at 1700 cm^{-1} is assigned to C = O stretching vibrations of ketones, aldehydes, lactones, or carboxyl groups. The band at 1600–1590 cm^{-1} is due to O–H bending. The band at 1423 cm^{-1} is attribuated to phenolic –OH and –C = O stretching of carboxylate. The band at 1372 cm^{-1} is characteristic of –COO groups. Moreover, a very strong band at 1032 cm^{-1} could be assigned to = C–O–C, P = O stretching, or P–OH stretching. While, the bands at 898, 830, and 700 cm^{-1} could be assigned to out-of-plane deformation mode of N-containing bioligands [6].

IR spectrum of red clay relates the presence of elongation vibrations relating to calcite (1442.5; 874 and 726 cm^{-1}), quartz (798 and 778 cm^{-1}), and calcite and dolomite. It shows also the presence of the OH hydroxyl group vibration of hydrated aluminosilicates structure, hydrated water molecules and poorly crystallized kaolinite (3698.5), of the water OH valence band from hydration of clay minerals (3453.6), OH valence bands of chlorite (3572 and 3498), characteristic bands of dolomite, associated with the valence vibration of the $CO_3{}^{2-}$ group (2526, 1817, 1450, 880 and 712) and elongation vibration bands of the calcite CO group.

The spectrum relating to DA is characteristic of apatite and more particularly carbonate fluoroapatite type B. In fact, the presence of absorption bands located at 1455 and 1430 cm^{-1} is shown on the spectrum. These positions are comparable to those observed in the case of carbonated phosphocalcic fluoroapatites in type B sites, prepared at boiling [12]. In addition, the IR spectra for the phosphate bands and bands around 780–800 cm^{-1} could come from the vibration of the silicate groups. A broad band in the 1620 cm^{-1} and 3460 cm^{-1} ranges is characteristic of the symmetrical valence vibrations of the OH- ions.

While the shoulder observed around 960 cm^{-1} is attributable to sulfate ions in the PG structure, these bands could be considered as evidence for the adsorption of the dyes on the RC, the DA, and PG.

3.1.2 Scanning Electron Microscopy (SEM)

Scanning electron microscopy images of *D. harra* and *G. coronaria L.* are presented in Fig. 2 and raw and treated sawdust of acacia are shown in Fig. 3. Those of the dust of apatite DA, phosphogypsum PG, and Red Clay RC are shown in Fig. 4.

Figure 2 shows that both the materials are characterized by a smooth and homogeneous surface. The surface of *D. Harra* does not have well-defined pores. However,

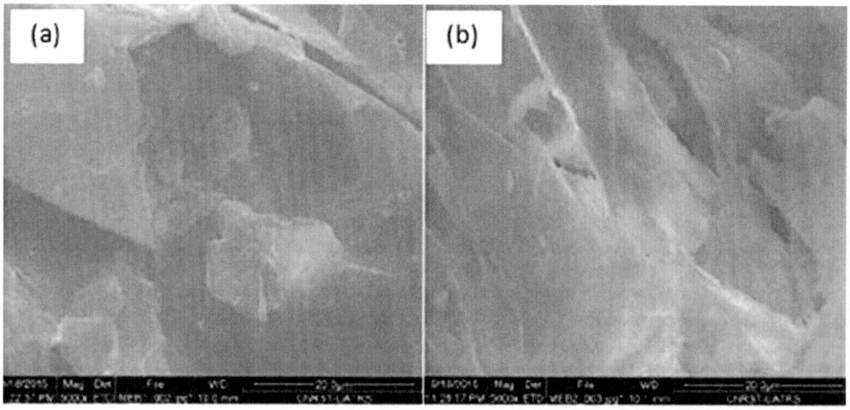

Fig. 2 SEM Images: **a** *D. harra*, **b** *G. coronaria L.*

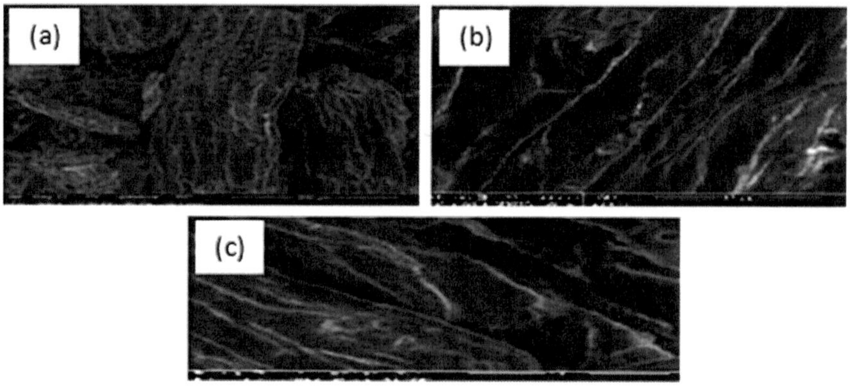

Fig. 3 SEM Images: **a** Raw-A, **b** A-HCl, **c** A-NaOH

Fig. 4 SEM images of: **a** PG, **b** DA, and **c** RC

the *G. Coronaria L.* fragment is formed by a smooth surface made up of some cavities. These results are in agreement with the low specific surface. Figure 3 indicates that the raw sawdust acacia contains many pores corresponding to the cellulose microfibrils concentrated inside the fiber. After the treatment process and due to the reaction between the raw material and the treatment agent, new pores of different sizes and cavities are formed. The external walls of the acacia fiber are composed mostly of hemicelluloses, and the interfiber joining lamellar containing the lignin. The last one binds the wood fibers together, while, the hemicelluloses join the cellulose and the lignins.

It can be deduced from Fig. 4 that the red clay has a lamellar structure confirmed by the stacking of layers, particles of irregular or rounded shapes, and pores in the form of holes.

The morphological investigation of Fig. 3 has revealed also that the AD has a beige color, including essentially phosphate particles of various irregular or rounded shapes, quartz grains, organic debris as well as bone debris. The sawn nodules on the left reveal a concentric structure, underlined by the variations in the concentration of organic matter. The grain structure results in a lower porosity and a total absence of honeycombing. These two characteristics provide for low specific surfaces.

The SEM images relating to PG show that the crystals are formed of fine fibrous and lamellar needles or in the form of a platelets stack, of yellowish beige color. These platelets are characterized by tabular facies with poorly defined edges and a very disturbed surface condition. The sulfated particles have a statistical distribution and almost tight dimensions which explains the low specific surface of the crystal.

3.1.3 Development of Surface Area

The calculated values of biosorbents' surface area are about 698.91 m^2/g for A-NaOH and 523.51 m^2/g for A-HCl. Hence, the NaOH treatment provides a large surface area for sawdust acacia compared to the HCl. This result suggests a difference in the sorption capacities of both the biosorbents.

The specific surface area of the *D. Harra* and *G. coronaria L.* biosorbents was 1.612 and 1.741 m^2/g respectively. This small surface area indicated that these biosorbents do not have any defined holes on their morphology.

The specific surface of DA, PG, and RC are 13, 16 and 40 m^2/g respectively. The specific surface of the PG is very weak and justifiable by the nature of its crystal lattice which consists of $CaSO_4$ sheets; between them water molecules are attached.

3.1.4 X-Ray Diffraction (XRD)

The X-ray diffractogram of the three samples including the raw and both the chemically treated acacia sawdust with HCl and NaOH are shown in Fig. 5.

The diffraction intensity increased after the chemical treatment of the acacia sawdust. These results suggest an increase in the crystallinity index according to

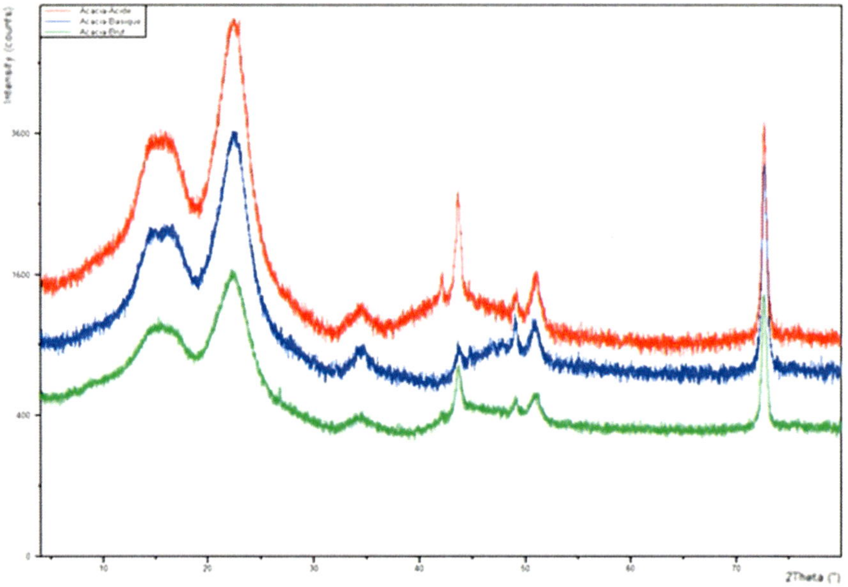

Fig. 5 XRD diagram of raw acacia, A-HCl, and A-NaOH

the chemical treatment applied to the sawdust that are due to the improvement in the order of crystallites. In fact, the degree of crystallinity of the cellulose is higher in the treated samples than in the raw sawdust, due to the reduction of hemicelluloses during chemical treatment. While, the treatment of acacia sawdust with HCl has an effect on increasing the crystallinity of the material, which is higher than that of acacia sawdust treated by NaOH. These results are consistent with those of Alemdar and Sain who made similar studies (Alemdar A. and Sain, 2008) on untreated and chemically treated Scots pine sawdust.

According to the literature [44], the mineral species that constitute RC adsorbent are illite, kaolinite, quartz, kaolinite, quartz and illite, calcite and dolomite respectively, shown by intense peaks 9.99 Å, 7.16 Å, 5.02 Å, 4.47 Å, 4.25 Å, 3.579 Å, 3.34 Å, 3.03 Å et 2.9 Å. The chemical composition of red clay is confirmed by the dolomite with the chemical formula $CaMg(CO_3)_2$ and kaolinite with the formula of $Al_2Si_2O_5(OH)_4$.

The DA adsorbent also crystallizes in the hexagonal system. It consists of a carbonate fluoroapatite, as the main constituent, large quantities of calcite, and a quantity of quartz. In the case of PG adsorbent, the main diffraction lines were easily indexed, in comparison with those of calcium sulfate dihydrate (Fig. 6). They are indexable in a monoclinic. The crystalline parameters of all phases were then calculated and compared to ASTM files, and the results were almost similar.

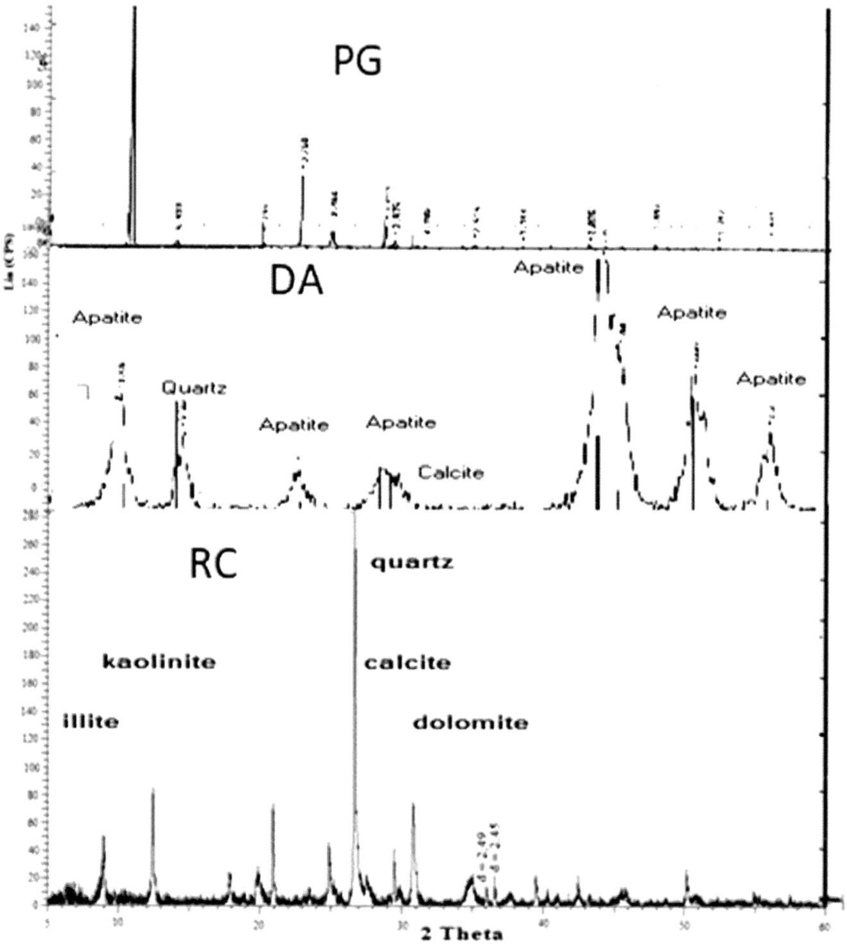

Fig. 6 Diffractogramms of DA, PG, RC adsorbents

3.2 Sorption Studies

3.2.1 Adsorption Kinetic

Adsorption rate kinetic gives important information for studying batch sorption. It demonstrates the rate of the reaction making it possible to determine the contact time involved in order to acquire the equilibrium of the reaction. The adsorption was rapid at the first period of the process, and after that the rate of adsorption becomes slower and stagnates with the increase in contact time.

The kinetic sorption of the MB dye onto A-HCl, A-NaOH, *D. harra*, *G. coronaria L.*, RC, DA, and PG is presented in Fig. 7. It could be seen for A-NaOH and also for A-HCl that the biosorption of methylene blue is rapid; the equilibrium is reached

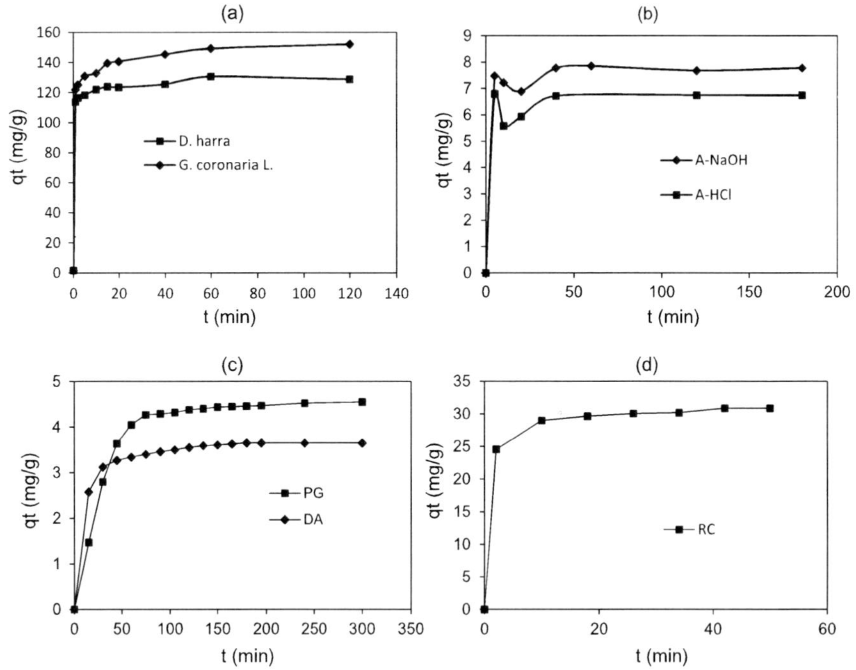

Fig. 7 Adsorption Kinetic of MB onto *D. harra* and *G. coronaria L.* (**a**), A-NaOH and A-HCl (**b**), PG and DA (**c**). RC (**d**), $C_{0(MB)} = 100$ mg/L, R = 0,5 g/L, T = 25 °C for *D. harra* and *G. coronaria L.*, A-NaOH and A-HCl C (MB):10 mg/L, adsorbent: 2 g/L of PG and DA, 400 mg/L of RC

after 90 min. In the case of *D. harra* and *G. coronaria L.*, it is reached after 60 min. Whereas, 50 min and 260 min are the time necessary to reach equilibrium for RC and for the PG and DA respectively.

In fact, the kinetic adsorption was initially fast for all the adsorbents. All the kinetic curves have the same shape and consist of two steps: the first for a rapid increase in the adsorbed quantity of dyes. A second showing a less rapid evolution, which ends with stabilization and saturation of the materials (adsorbed concentration at equilibrium). The second step is probably due to the intra-particle diffusion phenomena of the chromophoric groups of the dyes inside the pores of the adsorbents.

At the beginning, the increase in the amount of adsorbed dye could be due to the rise in the number of vacant sites available (Fig. 7).

For characterizing the kinetics involved in the adsorption process, pseudo-first-order and pseudo-second-order rate equations were applied, and the kinetic data were evaluated through the regression coefficient (r^2) and the amount of dye biosorbed per unit weight of the biosorbent.

The kinetic model fitting curves and the comparison of experimental and calculated sorption capacity (q) values can be applied to deduce the suitable kinetic models.

Also, the obtained correlation coefficient (r^2) is considered a main factor to acquire the adequate kinetic models and describe the sorption kinetics.

- *Pseudo-first-order kinetic model*

The pseudo-first-order rate of Lagergren is based on solid capacity. This model considers that the rate occupation of sorption sites are proportional to the number of unoccupied sites. It is one of the most mainly applied models for the sorption of a solute from liquid solution. It is generally established following the Eq. (3) [30]:

$$\log(q_e - q_t) = \log q_e - \frac{k_1}{2,303}t \tag{3}$$

where,

k_1 first-order reaction rate constant for MB adsorption on adsorbents (min^{-1}).
q_e amount of MB adsorbed at equilibrium (mg/g).
q_t amount of MB adsorbed at time t (mg/g).
t contact time (min).

This model makes it possible to describe the phenomena that happen during the first minutes of the adsorption process.

- *Pseudo-second-order kinetic model*

The pseudo-second-order model proposed by [24] is based on the assumption that the sorption follows second-order chemisorption. The pseudo-second-order modèl (PSO) is represented by Eq. (4):

$$\frac{t}{q_t} = \frac{1}{k_2 q_e^2} - \frac{1}{q_e}t \tag{4}$$

k_2 second-order reaction rate constant of MB sorption on adsorbents (g/mg/min).
q_e amount adsorbed at equilibrium (mg/g).
qt amount adsorbed at time t (mg/g).
t contact time (min).

By plotting t/qt = f (t), we get a line which gives k_2 and q_e.

The kinetic parameters of both the kinetic models are listed in Table 2.

From the results of Table 2, the correlation coefficients of the pseudo-first-order model obtained for MB biosorption onto A-NaOH and A-HCl, *D. harra* and *G. coronaria L.* are greater than those of the pseudo-second-order model. The theoretical qe biosorption capacities gave acceptable values in comparision to the experimental ones for the MB dye in the pseudo-first-order model. Hence, the reaction involved in this biosorption process is adequate to the pseudo-first-order kinetic model. Therefore, this model can describe more the sorption process of the MB onto these biosorbents. It can be seen that the removal of MB dye onto the studied biosorbents may be

Table 2 Kinetics parameters of pseudo-first-order and the pseudo-second-order models for the MB adsorption onto DA, PG, RC, A-NaOH, A-HCl, *D. Harra*, *G.coronaria. L*

	Pseudo-first-order				Pseudo-second-order		
	k_1 (min^{-1})	q_e($mg.g^{-1}$)	R^2	Q_{exp} (mg/g)	k_2($g.mg^{-1}$ min^{-1})	q_e($mg.g^{-1}$)	R^2
DA	0.023	2.917	0.980	3.650	0.039	3.750	0.990
PG	0.019	5.320	0.970	4.550	0.019	4.680	0.990
RC	0.100	10.200	0.900	25.350	0.048	25.420	0.990
A-NaOH	0.044	227.38	0.990	223.400	−1.008	187.43	0.740
A-HCl	0.030	221.77	0.990	209.00	1.194	163.92	0.980
D. Harra	2.319	124.713	0.985	131.163	2.702	123.36	0.977
G.Coronaria. L	1.756	142.046	0.961	154.186	3.149	139.099	0.938

done with a physical sorption. Also, this result suggests that the biosorption capacity may be due to the higher driving force making fast transfer of MB molecules to the surface of the biosorbent particles. Also, this sorption process can be attributed to the availability of the uncovered surface area and the remaining active sites on these adsorbents.

The results of the other type of adsorbents—RC, DA, and PG—are represented in Table 2. It appears that the values of R^2 are very higher and are all in the order of 0.99 in the case of the pseudo-second-order model. The calculated adsorption capacities of DA, PG, RC are 3.75 mg g^{-1}, 4.73 mg g^{-1}, and 25.42 mg g^{-1}, respectively, from the second-order model kinetic that are close to those found experimentally q_{exp}. Further, the q_{ecal} values obtained from the pseudo-first-order are different from the qe_{exp}. These results suggest that diffusion is the rate limiting step not only for the MB adsorption onto RC, DA, and PG but also the MB adsorption mechanism depended on both the MB adsorbate and the RC, DA, and PG adsorbents.

3.2.2 Effect of Adsorbent Dosage

Based on the variation of the amount of adsorbents, the data obtained are presented in Fig. 8. The figure indicates that the increase in biosorbent dosage resulted in a rise of the biosorption yield.

For *D. harra* and *G. coronaria L.*, the MB biosorption yield increased from 67.42 to 83.58% and from 46.48 to 86.82% when the biosorbent dosage was increased from 0.25 to 1 g/L, respectively. The effective biosorbent dosage of acacia shows a very rapid and a strong discoloration until 80% onto A-HCl and 90% onto A-NaOH using 0.5 g/L of these biosorbents. For RC, DA, and PG, MB dye removal increased from 6 to 23% for DA, 25 to 79% for PG, and from 61 to 92% for RC, by increasing the adsorbent dose from 1 to 7 g/L for DA and PG and from 100 to 700 mg/L for RC. The highest MB dye removal (92%) was obtained at 400 mg/L of RC, 4 g/L for

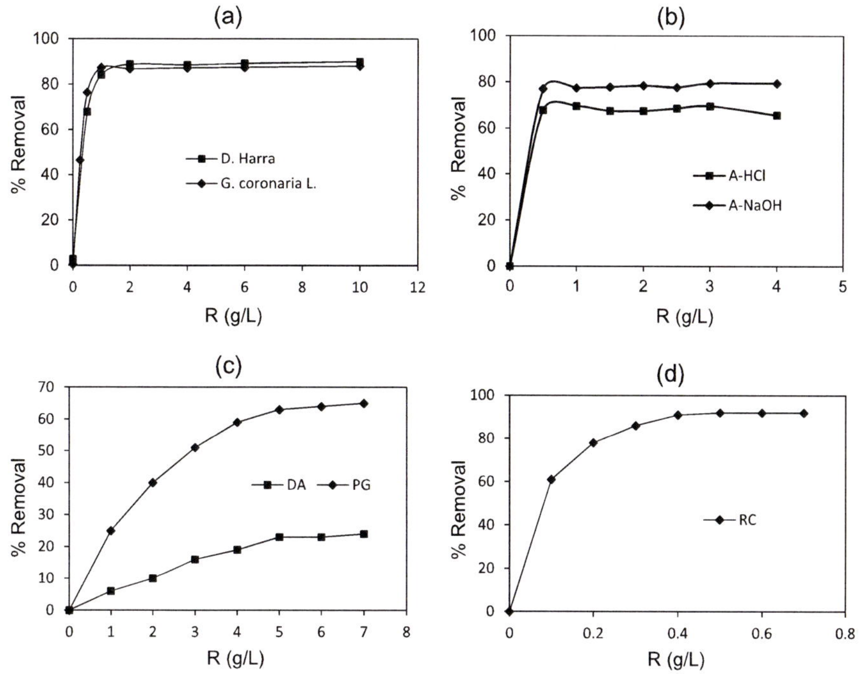

Fig. 8 Effect of adsorbent dosage on sorption of MB by *D. Harra* and *G. coronaria L.* (**a**); treated acacia sawdust; A-NaOH and A-HCl (**b**); $C_0 = 100$ mg/L, RC, AD, and PG, C (MB):10 mg/L, contact time = 120 min, pH = initial pH, T = 25 °C

DA, and 5 g/L for PG. Thus, a mass ratio of 0.5 g/L for A-NaOH and A-HCl, 1 g/L for *D. harra* and *G. coronaria L.*, 400 mg/L for RC, 4 g/L for DA, and 5 g/L for PG, respectively, are considered the optimal adsorbent dose with a higher percentage of removal of MB.

These results can be attributed to the higher surface area and/or active adsorption sites available in the surface of the studied materials for adsorption of MB dye molecules. Moreover, the increase in adsorbent quantities indicated no significant improvement in the MB dye removal. Similar trend was found by many authors [11, 58].

3.2.3 Adsorption Isotherms

Adsorption isotherms describe the state of the solute–surface interaction as well as the specific relation between the concentration of adsorbate and its degree of accumulation onto adsorbent surface at constant temperature.

The equilibrium biosorption capacity of adsorbents for MB removal increased with a rise in initial dye concentration. The increase in adsorbed amounts with concentration is probably due to a high driving force for mass transfer [33]. In fact, greater concentration in solution involved higher amount of dye molecules fixed at the surface of the adsorbent. Two isotherm models were applied to describe the adsorption experimental results: the Langmuir and the Freundlich models.

- *Langmuir isotherm model*

Langmuir proposed a theory to describe the adsorption of gas molecules onto metal surfaces [11]. Langmuir's model of adsorption predicts the existence of monolayer coverage of the adsorbate at the outer surface of the adsorbent. The isotherm equation further assumes that adsorption takes place at specific homogeneous sites within the adsorbent, which implies that all adsorption sites are identical and energetically equivalent.

The Langmuir isotherm can be expressed by the following form in Eq. (5):

$$\frac{Ce}{Q_e} = \frac{1}{K_L}\frac{q\max + C_e}{q\max} \tag{5}$$

where qe is the equilibrium dye concentration on the adsorbent (mg g^{-1}), Ce is the equilibrium dye concentration in solution (mg dm^{-3}), qmax is the monolayer capacity of the adsorbent (mg g^{-1}), and K_L is the Langmuir adsorption constant (dm^3 mg^{-1}).

- *Freundlich isotherm model*

The Freundlich isotherm is an empirical model of heterogeneous surface sorption with non-uniform distribution of heat sorption and affinities [18]. The form of the Freundlich equation can be stated as follows in Eq. (6):

$$Qe = K_F \frac{1}{nLn(C_e)} \tag{6}$$

where, K_F ($mg^{1-1/n}g^{-1}L^{1/n}$) and n are the Freundlich constants, n is the heterogeneity factor related to biosorption affinity and K_F is related to the biosorption capacity.

Adsorption parameters of Langmuir and Freundlich isotherm models of MB onto A-HCl, A-NaOH, DA, PG, RC, *G. coronaria L.*, and *D. Harra* for each model and correlation coefficients are summarized in Table 3.

This table implements that the Langmuir isotherm model investigates higher values of correlation coefficients in the adsorption of the MB dye onto DA, RC, *D. Harra*, *G. coronaria L.*, A-NaOH, and A-HCl adsorbents than the Freundlich isotherm model. Thus, the adsorption of MB onto these adsorbents is more fitted by Langmuir isotherm model. Whereas, the equilibrium data of PG are suitable to Freundlich model, with a higher correlation coefficient.

The equilibrium data of DA, RC, *D. Harra*, *G. coronaria L.*, A-NaOH, and A-HCl (Table 3, Fig. 9) can be explained by the fact that the increase in the dimensions of MB molecules due to the hydration, have prevented them from entering the pores

Table 3 Langmuir and Freundlich parameters for DA, PG, RC, *D. Harra*, *G. coronaria L.*, A-HCl, and A-NaOH adsorbents

Isotherms	Langmuir			Freundlich		
	Q_{max}	K_l	R^2	n	K_f	R^2
DA	16.36	0.307	0.999	1.297	3.644	0.995
PG	12.82	0.012	0.970	1.083	0.175	0.990
RC	50.29	3.26	0.990	6.622	36.59	0.800
D. Harra	161.29	0.08	0.990	5.319	64.43	0.880
G.coronaria L	185.59	0.308	0.980	3.333	55.95	0.720
A. NaOH	256.41	0.039	0.980	3.22	28.64	0.880
A. HCl	230.65	0.01	0.990	2.222	12.2	0.990

of the adsorbents, so that they are adsorbed on their external surfaces. It indicates also that the formation of monolayer takes place on the suface of the adsorbents, showing that only one dye molecule could be adsorbed on one site of adsorption.

While, the PG results following Freundlich model indicates a non-ideal and multilayer sorption on heterogeneous surfaces. For the Freundlich model, the values of $n > 1$ indicated a favorable biosorption under experimental conditions of MB dye sorption onto all the biosorbents.

It could be seen that the sorption capacities of all studying adsorbents are in the order of A-NaOH > A-HCl > *G.Coronaria L.* > *D. Harra* > RC > DA > PG. This result could be due to the higher surface area of A-NaOH than other biosorbents, where the NaOH increases the dissociated groups and creates well-developed pores on the surface of the material. It could be seen that the materials having greater specific surfaces have a higher sorption capacity unlike those with lower surfaces under the same conditions of experience. Then, the higher MB sorption onto adsorbent materials could be attributed to the nature of the interaction between each adsorbate and sorbent.

The maximum Langmuir adsorption capacities were 267.04, 230.65, 185.59, 161.29, 50.29, 12.82, 16.36 mg/g in the case of MB, respectively, for A-NaOH, A-HCl, *G. coronaria L.*, *D. Harra*, RC, DA, and PG. The adsorption capacities for MB dye were compared to the researches reported in literature of several low-cost sorbents as represented in Table 4. It could be seen that the experimental data of these already studied materials were higher than those of the most corresponding adsorbents in the literature. The adsorption capacities of these adsorbents can be attributed to the surface area of each material.

After studying the adsorption efficiencies of different materials as adsorbents and biosorbents of MB dye, it could be seen that the chemically treated sawdust of acacia tree has a greatest sorption capacity to remove MB dye from aqueous solutions. For that, a real final effluent of a textile industry was treated by sorption on both acidic and basic biosorbent based sawdust acacia.

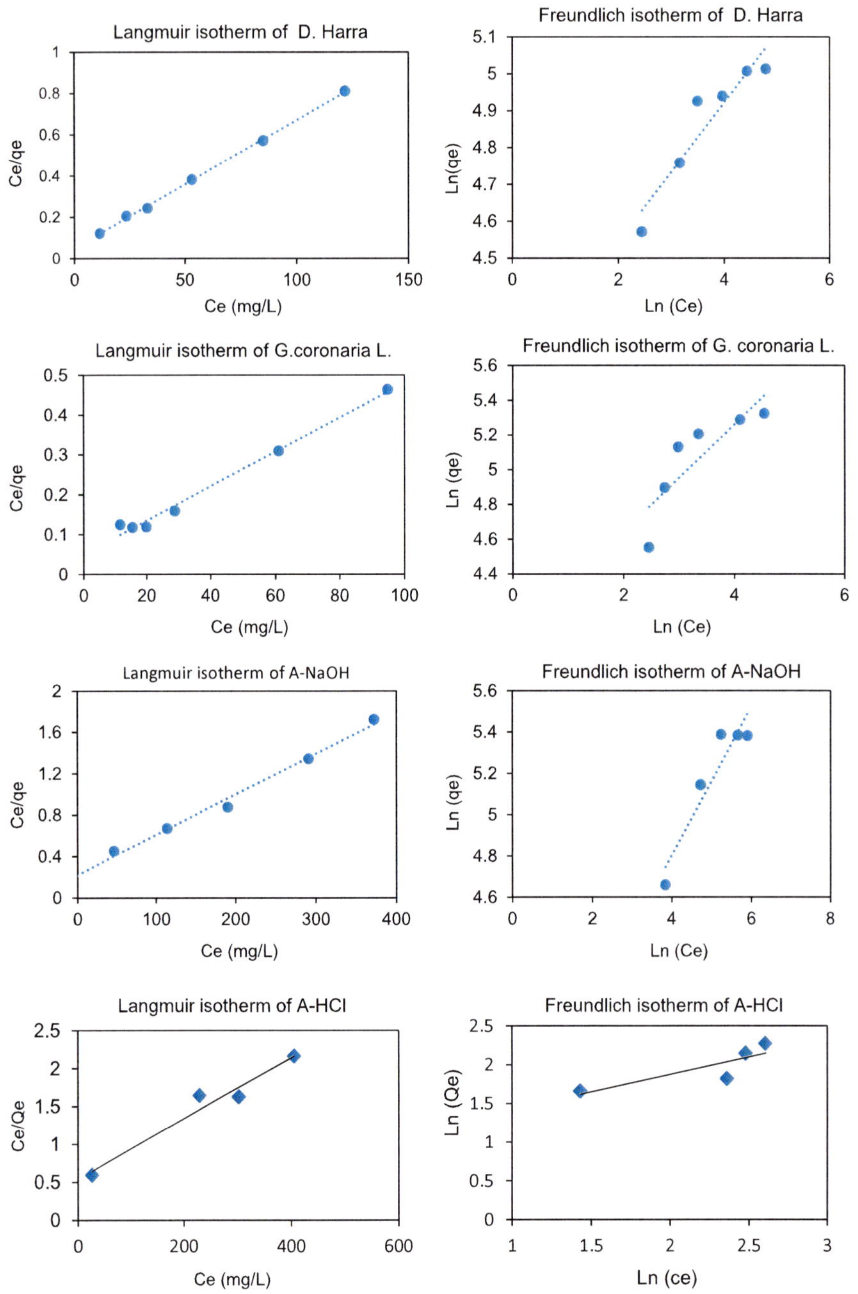

Fig. 9 Adsorption isotherm models of dye MB onto *D. Harra*, *G. coronaria L.*, A-NaOH, DA, PG, and RC adsorbents

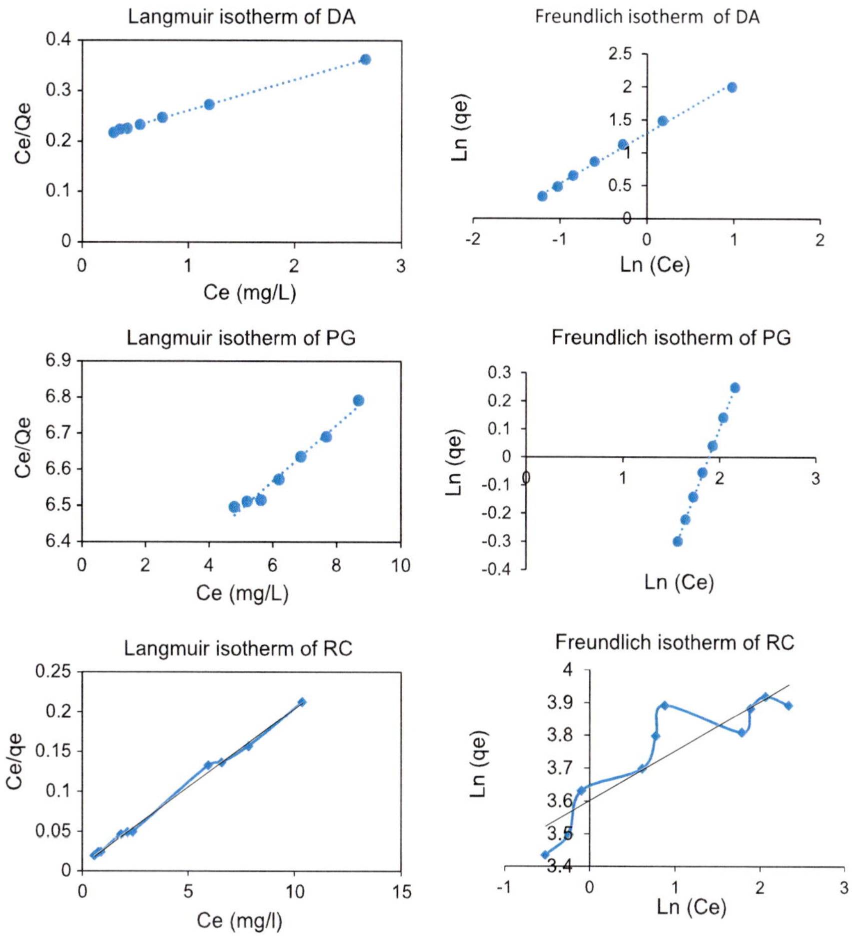

Fig. 9 (continued)

3.3 *Industrial Effluent Biosorption*

3.3.1 Physicochemical Characterization of the Studying Effluent

The average values of physicochemical characterization results of the studying effluent are established in Table 5 before biosorption. It appears that the average values of samples exhibit a temperature and a conductivity less than the standard values [13]. Hence, they have higher values for the other parameters.

The metallic characterization results of the studying effluents listed in Table 6 provides a high concentration of chromium 0.908 mg/L. It may be seen that it was the origin of the toxicity of different dyes applied in the treatment of jean, and therefore, its value was higher to standards of recommended rejects [13].

Table 4 MB adsorption comparison with differents adsorbents

Adsorbents	Q(MB) (mg/g)	References
Moroccan red clay from oued laou	18.83	[11]
Natural apatite	16.28	Our study
Phosphogypsum	13.40	Our study
Red clay of Safi	50.29	Our study
Natural clay	62.50	[44]
Algerian Kaolin	52.76	[40]
M. Bacillus subtilis	59 ± 0.60	[59]
Coconut coir dust	29.50	[51]
Carica papaya wood	32.25	[48]
Cotton waste	277.77	[39]
Coffee husks	90.10	[43]
D. Harra	161.29	[56]
G. coronaria L.	185.59	[56]
A-NaOH	256.41	[57]
A-HCl	230.65	[57]

Table 5 Results of the physicochemical characterization of the industrial effluent

	Temperature (°C)	pH	Conductivity (µs/cm)	DCO (mg O_2/L)	DBO_5 (mg O_2/L)	NTK (mg/L)	P (mg/L)	MES (mg/L)
Final discharge	27.13	9.01	186	658.36	336.3	126.49	2.4152	2194.6
Discharged Water Standards	30	5.5–8.5	2700	120	40	40	2	30

Table 6 Results of the metallic characterization

Heavy metals (mg/L)	Cd	Cr	Cu	Ni	Zn
Rejet final	0.0482	0.908	< 0.01	< 0.01	0.3021
Discharged Water Standards	0.2	0.5	3	5	5

3.3.2 Results and Characterization of the Industrial Effluent After Biosorption Studies

The kinetic sorption of the polluted charge allows decolorization rates of the industrial effluent of 77% and 98% for A-HCl and A-NaOH, respectively, with an optimal mass of 4 g/L [57].

Also, results indicate that the decolorization rate of the effluent adsorbed onto A-NaOH is more significant than that treated by A-HCl, suggesting that the biosorbent treated by NaOH is composed of more active sites than that treated with HCl. Moreover, this higher A-NaOH sorption can be attributed to the large surface area which is directly related to the porosity of this adsorbent and the presence of many active sites on its surface.

After biosorption and biological treatment elaborated by the textile industry, the obtained results of the industrial effluent characterization are given in Fig. 10.

This figure indicates that the treatment of industrial effluent by the biosorption on chemically treated acacia sawdust or by biological treatment carried out by the industry itself remains insufficient. Hence, the combination of both the treatments was recommended for better results since the biological treatment gives better results in terms of biological parameters (COD and BOD_5); while, the physicochemical treatment implements a good result, it treats effectively the color, the MES, and the odor of the studied effluent.

The biosorption wastewater of this study was compared to the previous records of various low-cost biosorbents as shown in Table 7.

It can be observed that the experimental data of A-NaOH and A-HCl were found among the hightest than those of the most corresponding adsorbents in the literature to remove color in real effluents.

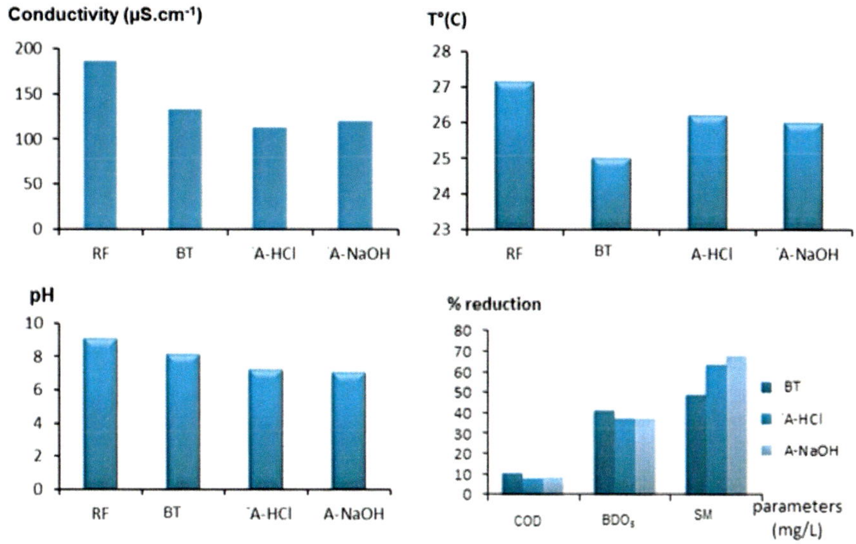

Fig. 10 Characterization of the industrial effluent after treatment

Table 7 Wastewater adsorption comparison with differents adsorbents

Adorbents/biosorbents	DCO removal percentage %	Abs removal percentage %	References
Arg.r	92	98	[9]
A-HCl	10	98	[57]
A-NaOH	8	77	[57]
Fungi	58	94	[45]
Moroccan pozzolana	94.5	Colorless	[55]
Activated carbon	45	94	[22]

4 Conclusion

In this work, chemically treated acacia sawdust, *Diplotaxis harra (D. harra)* and *Glebionis coronaria L. (G. coronaria L.)*, Dust Apatite (DA), Phosphogypsum (PG), and raw clay (RC) were used as low-cost biosorbents/adsorbents for the removal of methylene blue from aqueous solutions, and to find the optimal parameters. The sorption ability of these materials was investigated using kinetic and equilibrium studies and the effect of dosage adsorbents.

The biosorption of MB increases gradually up to 0.5 g/L of treated acacia sawdust, 400 mg/L of RC, 4 g/L of DA, and 5 g/L of PG. A further increase in biosorbent dosage beyond these values exhibited no extra improvement in MB removal.

Further, sorption experiments release that the sorption process was very rapid; the equilibrium time was obtained at 90, 60, and 50 min for *D. harra*, *G. coronaria L.,* and RC respectively and 260 min for PG and DA. The biosorption kinetic data fitted well to the pseudo-first-order kinetic model for chemically treated acacia sawdust, *D. harra* and *G. coronaria L.*, whereas, it follows to pseudo-second-order for Dust Apatite (DA), Phosphogypsum (PG), and Raw clay (RC).

The equilibrium data shows that the adsorption of MB onto A-NaOH, A-HCl, *D. Harra*, *G. coronaria L.*, RC, and DA was more fitted by Langmuir isotherm model. While, the equilibrium data of PG are suitable to Freundlich model. The adsorption equilibrium data of the methylene blue dye onto all the adsorbents and biosorbents were best fitted according to the Langmuir isotherm model. The Langmiur adsorption capacities are 267.04, 230.76 /g, 185.59, 258.76, 50.29, 16.36 and 12.82 mg/g onto basic sawdust acacia, acidic sawdust acacia, *D.Harra*, *G. coronaria L.*, RC, DA, and PG, respectively.

Under the experimental conditions, DA and PG did not give satisfactory results to remove dyes. While, the A-NaOH gives the most effective results than A-HCl, and than all other studied adsorbents. As a result, the adsorption process combined with a biological treatment investigates a good result based on physicochemical characteristics study.

Therefore, and for more enhanced adsorption results, a real final effluent of textile industry was treated onto A-NaOH and A-HCl that have been good sorption capacities. The kinetic sorption of the polluted charge allows decolorization rates of the industrial effluent of 77% and 98% for A-HCl and A-NaOH, respectively, with an optimal mass of 4 g/L. The higher removal of polluted charges has occurred through A-NaOH.

In prospect, for the other adsorbents—RC, DA, and PG—a pretreatment will be recommended before using them or mixing them with other high specific surface area co-products to increase their adsorption capacities and enhancing their properties to investigate a good sorption in the industrial scale.

Acknowledgements The authors thank the support from Engineering Laboratory of Electrochemistry, Modeling and Environment. Faculty of Sciences Dhar EL Mahraz, Sidi Mohamed Ben Abdallah University, Sultan Moulay Slimane University of Beni Mellal, Research Group in Environmental Sciences and Applied Materials (SEMA), FP Khouribga and Superior school of technologie, Dar Si Aissa Road, BP-89, Safi, Morocco.

References

1. Alemdar A, Sain M (2008) Isolation and characterization of nanofibers from agricultural residues—wheat straw and soy hulls. J Biores Technol 99(6):1664–1671
2. Aouni A, Fersi C, Dhahbi M (2014) Performance evaluation of direct nanofiltration process to fouling by treating rinsing-bath effluents for water reuse. Desalin Water Treat 52:1770–1785
3. Arulkumar M, Sathishkumar P, Palvannan T (2011) Optimization of Orange G dye adsorption by activated carbon of Thespesia populnea pods using response surface methodology. J Hazard Mater 186:827–834
4. Badi MKh, Samadi M, Asadian E, Zhou Y, Dong C, Zhang J, Moshfegh AZ (2019) Well designed Ag/ZnO/3D graphene structure for dye removal: Adsorption, photocatalysis and physical separation capabilities. J Colloid Interface Sci 537:66–78
5. Barka N, Qourzal S, Assabbane A, Nounah A, Ait-Ichou Y (2011) Photocatalytic degradation of patent blue V by supported TiO_2: kinetics, mineralization, and reaction pathway. Chem Eng Commun 198:1233–1243
6. Barka N, Abdennouri M, El Makhfouk M, Qourzal S (2013) Biosorption characteristics of cadmium and lead onto eco-friendly dried cactus (Opuntia ficus indica) cladodes. J Environ Chem Eng 1:144–149
7. Bharat NP, Naik DB, Shrivastava VS (2011) Photocatalytic degradation of hazardous Ponceau-S dye from industrial wastewater using nanosized niobium pentoxide with carbon. J Desalination 269(1–3):276–283
8. Bennani KA, Mounir B, Hachkar M, Bakasse M, Rais Z, Yaacoubi A (2010) Removal of the basic dye "Methylene blue" in aqueous solution using clay of Safi. J Water Sci 23(4):375–388
9. Bennani KA, Mounir B, Hachkar M, Bakasse M, Rais Z, Yaacoubi A (2017) Adsorption of BR46 dye using raw and purified clay. JOWSET (02), 233–240
10. Bennani KA, Mounir B, Hachkar M, Bakasse M, Yaacoubi A (2015) Adsorption of cationic dyes onto Moroccan clay: Application for industrial wastewater treatment. J Mater Environ Sci 6(9):2483–2500
11. Bentahar Y, Draoui K, Hurel C, Ajouyed O, Khairoun S, Marmier N (2019) Physico-chemical characterization and valorization of swelling and nonswelling Moroccan clays in basic dye removal from aqueous solutions. J Afric Earth Sci 154, 80–88
12. Boyer A (2014) Synthesis, characterization and biological evaluation of carbo-silicate phosphocalcicapatites; Mines National School of Saint-Etienne, Thesis memory
13. Official Bulletin, Official translation edition, Kingdom of Morocco (2013) 6202

14. Elmoubarki R, Mahjoubi FZ, Tounsadi H, Moustadraf J, Abdennouri M, Zouhri A, El Albani A, Barka N (2015) Adsorption of textile dyes on raw and decanted Moroccan clays: kinetics, equilibrium and thermodynamics. Water Res Ind 9:16–29
15. El Boujaady H, Mourabet M, El Rhilassi A, Bennani-Ziatni M, El Hamri R, Taitai A (2017) Interaction of adsorption of reactive yellow 4 from aqueous solutions onto synthesized calcium phosphate. J Saudi Chem Soc 21(1):S94–S100
16. Elmerzouki K, Khalidi A, Abdelhedi R, Bimaghra I, Taxil P, Lafage B, Savall A (2012) Removal of Cr(VI) and Pb(II) from aqueous solution by activated sawdust. Phys Chem News 64:76–86
17. Falil F, Allam F, Gourich B, Vial Ch, Audonnet F (2016) Adsorption of Astrazon Orange G onto natural Moroccan phosphate rock: a mechanistic study. J Envir Chem Engin 4:2556–2564
18. Freundlich H, Heller W (1939) The Adsorption ofcis- andtrans- Azobenzene. J Am Chem Soc 61(8):2228–2230
19. Gabriela CS, Virginia STC, Angela MF, Nathalia CP, Paulo Renato PP, Jorge LL (2014) A facile synthesis of Mn_3O_4/Fe_3O_4 superparamagnetic nanocomposites by chemical precipitation: characterization and application in dye degradation. Mater Res Bull 49:544–551
20. Gayathri PV, Yesodharan S, Yesodharan EP (2019) Microwave/Persulphate assisted ZnO mediated photocatalysis (MW/PS/UV/ZnO) as an efficient advanced oxidation process for the removal of RhB dye pollutant from water. J Environ Chem Eng 7(4):103122
21. Gholami-Borujeni F, Naddafi K, Nejatzade- BF (2013) Application of catalytic ozonation in treatment of dye from aquatic solutions. Desalin Water Treat 51:6545–6551
22. GilPavas E, Dobrosz-Gómez I, Miguel-Ángel Gómez- García (2020) Efficient treatment for textile wastewater through sequential electrocoagulation, electrochemical oxidation and adsorption processes: Optimization and toxicity assessment. J Electroanal Chem 878, 114578
23. Giwa A, Dindi A, Kujawa J (2018) Membrane bioreactors and electrochemical processes for treatment of wastewaters containing heavy metal ions, organics, micropollutants and dyes: recent developments. J Hazard Mater 370:172–195
24. Ho YS, Mckay G (1998) The kinetics of sorption of basic dyes from aqueous solution by sphagnum moss peat. Canad J Chem Engin 76(4):822–827
25. Issa AA, Al-DegsY S, Al-Ghouti MA, OlimatA A (2014) Studying competitive sorption behavior of methylene blue and malachite green using multivariate calibration. J Chem Eng 240:554–564
26. Jisha J, Raji CR, Johnly KJ, Steny PJ, Jency T (2020) Ion-exchange mediated removal of cationic dye-stuffs from water using ammonium phosphomolybdate. J Mat Chem Phys 242, 122488
27. Katheresan V, Kansedo J, Lau SY (2018) Efficiency of various recent wastewater dye removal methods: a review. J Envir Chem Engin 6(4):4676–4697
28. Kiayi Z, Lotfabad TB, Heidarinasab A, Shahcheraghi F (2019) Microbial degradation of azo dye carmoisine in aqueous medium using Saccharomyces cerevisiae ATCC 9763. J Hazard Mater 373:608–619
29. Lafi R, Gzara L, Lajimi RH, Hafiane A (2018) Treatment of textile wastewater by a hybrid ultrafiltration/electrodialysis process. Chem Eng Process-Process Intensif 132, 105–113
30. Lagergren S (1898) Zur theorie der sogenannten adsorption gel¨oster stoffe. Kungliga Svenska Vetenskapsakademiens, Handlingar 24(4):1–39
31. Langmuir (1916) The constitution and fundamental properties of solids and liquids: Part I: solids. J Americ Chem Soc 38(1):2221–2295
32. Leebana VJ, Santhanam H, Geetha K, Raj SA (2014) Biodegradation of direct golden yellow, a textile dye by Pseudomonas putida. Desalin Water Treat 39:1–9
33. Li K, Wang X (2009) Adsorptive removal of Pb(II) by activated carbon prepared from spartina alterniflora: equilibrium, kinetics and thermodynamics. Biores Technol 100:2810–2815
34. Li J, Li Y, Xiong Z, Yao G, Lai B (2019) The electrochemical advanced oxidation processes coupling of oxidants for organic pollutants degradation: a mini-review. Chin Chem Lett 30(12):2139–2146
35. Liina O, Iakov K, Sergei P (2020) Oxidation of reactive azo-dyes with pulsed corona discharge: Surface reaction enhancement. J Electrost 103, 103420

36. Machrouhi A, Farnane M, Elhalil A, Elmoubarki R, Abdennouri M, Qourzal S, Tounsadi H, Barka N (2017) Effectiveness of beetroot seeds and H_3PO_4 activated beetroot seeds for the removal of dyes from aqueous solutions. J Water Reuse Desalination 8:522–531. https://doi.org/10.2166/wrd.2017.034
37. Marty S (2014) Production of calcium oxide and Sulfur dioxide from the decomposition of phosphogypsum. Polytechnic School of Montreal, master's thesis
38. Mas Haris MRH, Sathasivam K (2009) The removal of methyl red from aqueous solution using banana pseudostem fibers. J Am Appl Sci 6:1690–1700
39. McKay G, Porter JF, Prasad G (1999) The removal of dye colors from aqueous solutions by adsorption on low-cost materials. Water Air Soil Pollut 114:423–438
40. Mouni L, Belkhiri L, Bollinger JC, Bouzaza A, Assadi A, Tirri A, Dahmoune F, Madani K, Remini H (2018) Removal of methylene blue from aqueous solutions byadsorption on kaolin: kinetic and equilibrium studies. Appl Clay Sci 153, 38–44
41. Mu B, Li W, Xu H, Xu L, Yang Y (2018) Freeze-extrusion for controllable assembly of 3-dimensional ultra-fine and amorphous fibrous matrices: potential applications in sorption. J Mater Chem A 6(22):10320–10330
42. Nippatla N, Philip L (2019) Electrocoagulation-floatation assisted pulsed power plasma technology for the complete mineralization of potentially toxic dyes and real textile wastewater. Process Saf Environ Prot 125, 143–156
43. Oliveira LS, Franca AS, Alves TM, Rocha SDF (2008) Evaluation of untreated coffee husks as potential biosorbents for treatment of dye contaminated waters. J Hazard Mater 155:507–512
44. Omer OS, Hussein MA, Hussein BH, Mgaidi A (2018) Adsorption thermodynamics of cationic dyes (methylene blue and crystal violet) to a natural clay mineral from aqueous solution between 293.15 and 323.15 K. Arab J Chem 11(5):615–623
45. Prigione V, Tigini V, Pezzella C, Anastasi A, Sannia G, Cristina VG (2008) Decolourisation and detoxification of textile effluents by fungal biosorption. Water Res 42:2911–2920
46. Qasim M, Badrel Zaman M, Darwish NN, Darwish NA, Hilal N (2019) Reverse osmosis desalination: a state-of-the-art review. Desalination 459, 59–104
47. Rais Z, Elhaji M, Taleb M, Zemzami M, Benabbou M (2015) Valorization of the apatitic Calcium Phosphate in the treatment of textile waste Effluent. Int J Mater Chem Phys 1:26–30
48. Rangabhashiyam S, Lata S, Balasubramanian P (2018) Biosorption characteristics of methylene blue and malachite green from simulated wastewater onto Carica papaya wood biosorbent. Surf Interfaces 10:197–215
49. Rodier J (2009) L'analyse de l'eau - 9'eme ´edition - Eaux naturelles, eaux r´esiduaires, eau de mer, Dunod, Paris, France, 9eme edition
50. Rosa JM, Fileti AMF, Tambourgi EB, Santana JCC (2015) Dyeing of cotton with reactive dyestuffs: the continuous reuse of textile wastewater effluent treated by ultraviolet/hydrogen peroxide homogeneous photocatalysis. J Clean Prod 90:60–65
51. Runping H, Yuanfeng W, Pan H, Jie S, Jian Y, Yongsen L (2006) Removal of methylene blue from aqueous solution by chaff in batch mode. J Hazard Mater 137:550–557
52. Sena M, Hicks A (2018) Life cycle assessment review of struvite precipitation in wastewater treatment. Resour Conserv Recycl 139:194–204
53. Shuang Z, Zhan W (2017) A loose nano-filtration membrane prepared by coating HPAN UF membrane with modified PEI for dye reuse and desalination. J Membr Sci 524:214–224
54. Saleem J, Shahid UB, Hijab M, Mackey H, McKay G (2019) Production and applications of activated carbons as adsorbents from olive stones 9(4):775–802
55. Titchou FE, Ait Akbour R, Assabbane A, Hamdani M (2020) Removal of cationic dye from aqueous solution using Moroccan pozzolana as adsorbent: isotherms, kinetic studies, and application on real textile wastewater treatment, Groundwater for sustainable development 11 100405
56. Tounsadi H, Khalidi A, Abdennouri M, Barka N (2015) Potential capability of natural biosorbents: diplotaxis harra and Glebionis coronaria L. on the removal efficiency of dyes from aqueous solutions. J Desalin Wat Treat 1072954

57. Tounsadi H, Metarfi Y, Barka N, Taleb M, Rais Z (2020) Removal of textile dyes by chemically treated sawdust of acacia: kinetic and equilibrium studies. Hindawi J Chem 7234218
58. Toor M, Jin B (2012) Adsorption characteristics, isotherm, kinetics, and diffusion of modified natural bentonite for removing diazo dye. J Chem Engin 187:79–88
59. Tural B, Ertas E, Enez B, Fincan SA, Tural S (2017) Preparation and characterization of a novel magnetic biosorbent functionalized with biomass of Bacillus Subtilis: kinetic and isotherm studies of biosorption processes in the removal of methylene blue. J Environ Chem Engin 5(5):4795–4802
60. Zahrim AY, Hilal N (2013) Treatment of highly concentrated dye solution by Coagulation/flocculation–sand filtration and nanofiltration. Water Res Ind 3:23–34
61. Zollinger H (1991) Color chemistry: synthesis, properties and applications of organic dyes and pigments. VCH, New York, NY

Ability to Remove Azo Dye from Textile Dyeing Wastewaters of Carbonaceous Materials Produced from Bamboo Leaves

Thi Thu Huong Tran, Ngoc Toan Vu, Thanh Nga Pham, and Xuan Tong Nguyen

Abstract Textile dyeing wastewaters is one of the most difficult to handle. The aim of this paper to investigate the removal capacity of two azo dyes, including Brilliant green and Alizarin red S from the aqueous solution by three types of activated carbon prepared from bamboo leaves AC30 (650 °C/30 min), AC45 (650 °C/45 min) and AC60 (650 °C/60 min). All three samples displayed the functional group represented by the azo dye removal capacity such as C=C; C–O–C, O–H and showed high C content (over 72%). With the highest BET surface area up to 108.9202 m^2/g, the AC60 material sample recorded the maximum efficiency of 100% at the reaction time of 30 min with the volume ratio of azo dye/distilled water of 2/18 (mL), pH 9, the absorbent amount of 0.5 and 2 g for Brilliant green and Alizarin red S, respectively. This study is an overview of the azo dye removal capacity in initial textile dyeing wastewater sample and compares the quality of AC60 material sample to four commercial activated carbon samples, including AC-R (Russian activated carbon), AC-C (Chinese activated carbon), AC-F (French activated carbon) and AC-Tra Bac (Vietnamese activated carbon from coconut shell). The analysis results also showed that the azo dye removal efficiency of the AC60 material synthesized from bamboo leaves is higher than some previously prepared material (AC-C material reached only 90.5 and 82.55% for Brilliant green and Alizarin red S, respectively) and achieved the maximum adsorption efficiency 100% after 30 min reaction time. These findings indicated that the removal efficiency for azo dye depends on experimental conditions

T. T. H. Tran
Faculty of Environment, Hanoi University of Mining and Geology, 18 Vien Street, Duc Thang Ward, Bac Tu Liem District, Hanoi, Vietnam

N. T. Vu
Institute of New Technology – Material, Institute of Military Science and Technology, 17 Hoang Sam Street, Cau Giay District, Hanoi, Vietnam

T. N. Pham
Faculty of Chemistry, Hanoi University of Education, 136 Xuan Thuy Street, Cau Giay District, Hanoi, Vietnam

X. T. Nguyen (✉)
Institute of Environmental Science, Engineering, and Management, Industrial University of Ho Chi Minh City, 12 Nguyen Van Bao street, Go Vap District, Ho Chi Minh City, Vietnam

© The Author(s), under exclusive license to Springer Nature Singapore Pte Ltd. 2021
S. S. Muthu and A. Khadir (eds.), *Novel Materials for Dye-containing Wastewater Treatment*, Sustainable Textiles: Production, Processing, Manufacturing & Chemistry,
https://doi.org/10.1007/978-981-16-2892-4_8

such as reaction time, pyrolysis temperature, pH... and the source of the raw materials also has a great influence on the carbonaceous material structure, determining the treatment efficiency as well as the removal capacity for pollutants.

Graphic Abstract

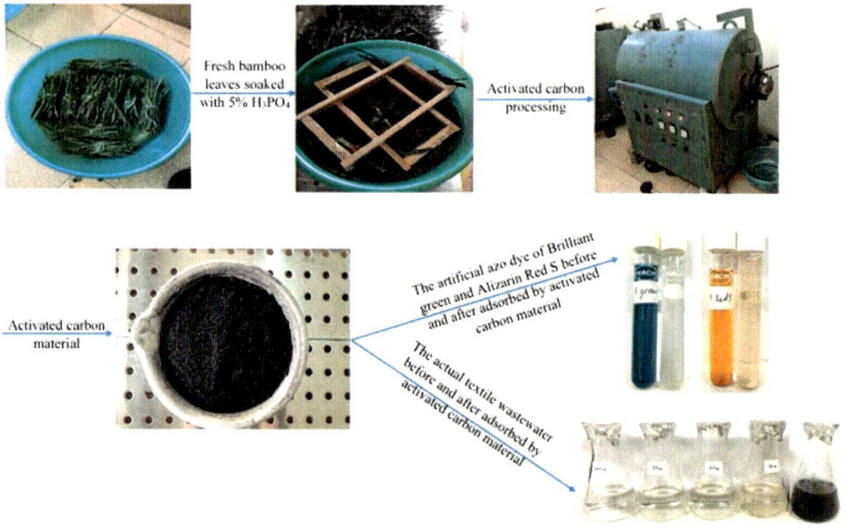

Keywords Activated carbon · Bamboo leaves · Azo dyes · Adsorption efficiency · Adsorption capacity · Brilliant green · Alizarin red S · Textile wastewater · Carbonaceous material · Characteristic material

1 Introduction

Activated carbon (AC) is well known as the most commonly used adsorbent in water and wastewater treatment. ACs characterized by their extended specific surface reactivity, good internal porous structure and high chemical, mechanical and thermal stability [15] which be used in a variety of applications such as removal of organic pollutants, heavy metals, medical application, catalysis, electrode materials in electrochemical devices [5]. In recent years, the adsorption techniques of ACs have been widely applied in water treatment due to fast adsorption kinetics, simplicity of design, initial cost, high removal efficiency, and insensitivity to toxic substances. However, the high cost of ACs makes it increasingly necessary to produce and regenerate [7, 39]. The carbonaceous material has a wide range of sources, including industrial waste [13], waste of biodiesel industry [36], agricultural by-products (pineapple waste [32]; soybeans, corn stalks, corn stalks [29], corncob [49]), livestock manure (poultry manure, poultry manure, poultry manure [29, 54]), vermicompost [51] sludge. The carbonaceous material itself has the advantages of large pores, high surface area, and variety surface functional groups. The variety surface functional groups with high porosity are considerably crucial in the adsorption process of dye, especially azo dyes. Azo dyes are widely used in various industries such as textiles, leather, plastics, and paper and manufacturing of inks, which account for more than 60% of total dyes.

Azo dyes characterized with the presence of one or more azo functional groups (R_1–N=N–R_2), where R_1 and R_2 are aromatic groups, can be substituted by some combinations of functional groups such as amino (–NH_2), chlorine (–Cl), hydroxyl (–OH), methyl (–CH_3), nitro (–NO_2), sulphonic acid and sodium salts (–SO_3Na) [8]. However, azo dyes as wastewater pollutants represent a big concern for the environment, and currently the most attractive issue is finding the methods to remove these substances from water. Most dyes are toxic, carcinogenic or mutagenic [20] and can pose a hazard to health, like causing chromosomal damage [2]; affecting the eyes and skin, damage to internal organs such as liver and kidney [3]; disrupting the photosynthesis process in water bodies [38]. Consequently, it is very essential to treat the textile wastewater or find the new material to absorb or support the azo dye removal from wastewater. Several techniques were investigated to remove the azo dye species from the wastewater [45], including chemical, physical and biological treatments or a combination of these processes such as: adsorption [10, 26], hydrogen peroxide [25], biological treatment [24], membrane filtration [28]. Among the physical processes, adsorption is considered an effective separation technique in terms of low cost, flexibility, simplicity of design, ease of operation, and sensitivity to toxic substances [1, 20].

Bamboo is a large, woody-grass member of the family of Bambusoideae encompassing about 1250 species within 75 genera worldwide [16]. The raw materials used in the paper industries, cottage industries, domestic commodities, board and charcoal are gradually being replaced by bamboo material [30]. Vietnam is the fourth country in the world in terms of area planted to bamboo and production of bamboo, primarily grown in Lam Dong, Da Lat. Vietnamese people use bamboo in many ways, such as sharpened bamboo trees to make weapons in the wars, and now, bamboo trees are used as materials to build houses, make furniture, and other useful everyday items baskets, chopsticks, fans.

Bamboo is considered one of the ideal materials for the twenty-first century because it is environmentally friendly. Nowadays, there has been increasing interest in the research of the production of low-cost activated carbon from bamboo. Many studies have been conducted to evaluate the ability of using bamboo in the synthesis of carbonaceous materials for the removal of dyes [14, 52]. In this study, the carbonaceous materials were synthesized from bamboo leaves at 650 °C with three ranges of activation time 30, 45, and 60 min. The adsorption capacity of the activated carbon material in current study was evaluated through the use of two industrial dyes (namely, Brilliant green and Alizarin red S). The characteristics and structure of two selected commercial azo dyes are shown in Table 1 and Fig. 1 below.

Various researches were carried out to remove these dyes from wastewater by the carbonaceous material synthesized from other residues or agriculture waste, such as: studies for the treatment of Naphthol B. green by the red mud [25], removal of Alizarin red S reviewed by the activated carbon/γ-Fe_2O_3 nano-composite [9, 23],

Table 1 Characteristics of two azo dyes used in the current study

Dye	Brilliant green	Alizarin red S
Molecular formulae	$C_{27}H_{33}N_2HO_4S$	$C_{14}H_7NaO_7S$
IUPAC name	4-{[4 (diethylamino)phenyl](phenyl)methylidene}-N,N-diethylcyclohexa-2,5-dien-1-iminium	1,2-dihydroxy-9,10-anthraquinonesulfonic acid sodium salt
Characteristic	It is a cationic dye. It causes an allergic reaction on the human body, eye irritation and even leads to blindness, directly affects in the liver and kidney failure, cancer-causing agent, and even mutations in human beings [27, 34]	It is a water-soluble, anionic anthraquinone dye and usually used as a staining agent in textile industries. This dye is also used to stain biological specimens such as mineralized bones invertebrate groups and tiny invertebrate embryos [42]

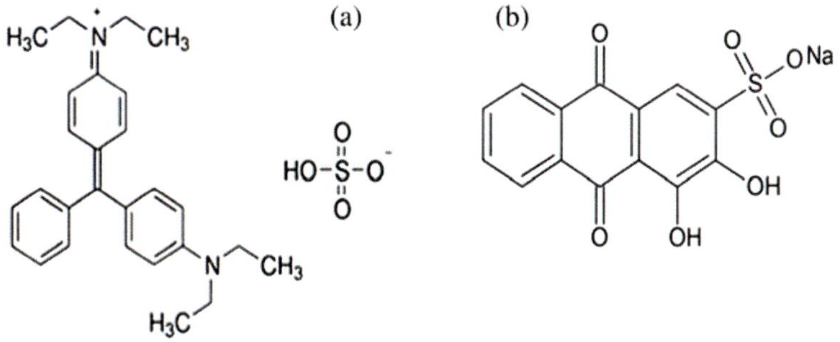

Fig. 1 Structure of two selected commercial azo dyes: **a** Brilliant green [27] and **b** Alizarin red S [42]

removal of Chrysoidine Y by the activated Sawdust [6, 35]. However, not much have been published on using ACs material obtained from bamboo leaves to adsorb these above dyes from industrial effluents. Therefore, this study was conducted to evaluate the ability to remove two azo dyes from the aqueous solution, including Brilliant green and Alirazin red S by carbonaceous materials produced from bamboo leaves. The study was designed to (1) assess the azo dye removal efficiency (according to the amount of the adsorbent material and the reaction time) of three activated carbon material and (2) examine the actual wastewater sample and (3) assess the quality of the synthesized material by comparing the adsorption efficiency to some commercial carbonaceous material.

2 Materials and Methods

2.1 Preparation of Activated Carbon from Bamboo Leaves

Fresh bamboo leaves as raw materials were collected at Lang area, Thach That District, Hanoi city, Vietnam and moved to laboratory, clean washed and naturally dried at room temperature for 24 h. Bamboo leaves were then chopped and soaked with 5% H_3PO_4 solution for 1–2 h, took out to drain (the aim of this technical step to clean impurities on bamboo leaves and prevented the natural burning under heating effect). Transfer the bamboo leaves to the tube containing samples of the oven and carbonized them at a temperature of 225–230 °C/4 h. Afterwards, down the temperature of the oven to 150 °C and took the carbonized bamboo leaves out of the oven. Then, put these bamboo leaves into distilled water to reduce heating, soon after took out the basket to drain naturally. After draining, put carbonized bamboo leaves into each inox tray in the drying oven, dried at 60 °C for 12 h. The final, transfer the carbonized bamboo leaves to each ceramic bowl to activate in the UAF furnace at temperature 650 °C for 30, 45, and 60 min, respectively. The washed material samples were dried at 105 °C for 12 h, finely ground and stored in ceramic boxes.

The yield of activated carbon in the current study calculated to be 38.71 [19]. The three material samples will be encoded as AC30, AC45, AC60 and used for further experiments. All the chemicals in this study supplied by Merck.

2.2 *Identification of Material Characteristics*

The activated carbon material also determined for its surface morphological characteristics and the chemical composition using a JSM 6380 Scanning Electron Microscope (SEM) and Energy-dispersive X-ray spectroscopy (EDX). The functional groups on the surface of each material sample were identified using the Fourier Transform Infrared spectroscopy (FTIR) TENSOR II instrument. The BET- and t-plot method was used to determine the total surface area, average pore radius and micropore volume of samples by Micromeritics Tristar 3000 instrument. Besides, the particle size is also determined to have an overview of the structure of the three activated carbon materials. All samples in this study were measured at the Institute of Tropical Technology, Vietnam Academy of Science and Technology and Hanoi National University of Education.

2.3 *Experimental Setup*

2.3.1 Preparation of Azo Dyes (Brilliant Green and Alizarine Red S)

The two azo dyes used in current study (Brilliant green and Alizarin red S) were supplied by Sigma–Aldrich. Their chemical structure is shown in Fig. 1. Weigh 0.01 g each compound Brilliant green and Alizarin red S separately, dissolve in 100 mL of distilled water, shake well and store in a dark bottle as a stock solution.

2.3.2 Experimental Steps

Our steps proceed very much in the same way follow what is indicated in Tran et al. [48]. Three activated carbon materials were evaluated for the adsorption capacity in batches in 100 mL flasks. At first, carried out experiments to select the optimization volume of azo dye in solution. 0.5 g activated carbon added into 20 mL solution containing azo dye (Brilliant green and Alizarin red S) and distilled water according to the ratio of $V_{\text{azo dye solution}}: V_{\text{distilled water}} = 2{:}18; 3{:}17; 4{:}16; 5{:}15$ and no pH adjustment. Then, shake the samples well by a horizontal shaker with speed of 150 r/m for 30, 60, 90, 120 min and measure the pH of the newly mixed samples. The filtered samples have analyzed the concentration of azo dye remaining by HPLC instrument at 360 nm (Brilliant green) and 260 nm (Alizarin red S) in Institute of Chemistry - Material, Institute of Military Science and Technology, Hanoi, Vietnam. The ratio of $V_{azo\ dye\ solution}$: $V_{distilled\ water}$ with the highest adsorption efficiency was selected to evaluate for further experiments.

Subsequently, 0.25; 0.5; 1; 2 g of three material samples added into every flask contains 20 mL solution with selected optimization ratio and continue to carry out experiments at 30, 60, 90, 120 min. Adsorption capacity q_e (mg/g) and adsorption efficiency H (%) at equilibrium time are calculated according to the formula of Tran et al. [48] as follows:

$$q_e = (C_o - C_e) \cdot V/m \ (mg/g)$$

$$H(\%) = (C_o - C_e)/C_o \cdot 100 \ (\%)$$

where q_e is the adsorption capacity at equilibrium (mg/g), C_o: initial concentration (mg/L), C_e: concentration at equilibrium (mg/L), V: volume of solution (L), m: mass of absorbent material (g).

In this study, V (the adsorption volume, 20 mL), C_o (the concentration of the initial adsorbent solution, mg/L), and m (the amount of the adsorbent, 0.25; 0.5; 1 and 2 g) were fixed. The ratio volume of azo solution/distilled water and adsorption time will be changed various from 2:18; 3:17; 4:16 và 5:15 mL and t = 30; 60; 90; 120 min, respectively.

The experimental procedure is summarized in Fig. 2 below. The newly synthesized activated carbon with highest adsorption capacity will be used to compare the quality with some available commercial products (from Russia, China, France and the Tra Bac of Vietnam). A textile wastewater sample was collected at the local textile factory in Long Bien, Hanoi and transfer to the laboratory. Analyze the initial

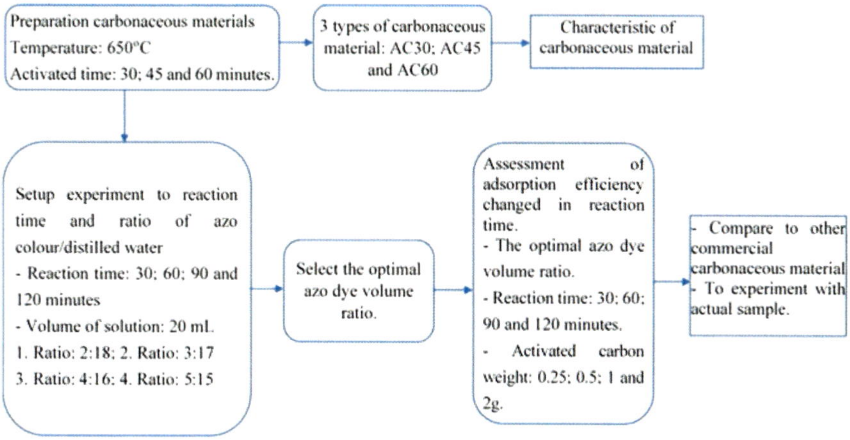

Fig. 2 Setup the experimental conditions

pollution concentration and test with the optimization conditions. Evaluate the azo dye adsorption capacity of the material.

2.4 *Statistical Analysis*

Statistical significance was analyzed through the use of Origin 2019 and GraphPad software.

3 Results and Discussion

3.1 *Characteristics of Activated Carbon Material*

In this study, the material characteristics of the three activated carbon samples is shown in Table 2 and Figs. 3, 4, 5. The C content of all material samples is relatively high, in which the AC60 sample has the highest C surface content of 80.18%,

Table 2 The physical and chemical properties of three activated carbon material

Code sample	Activated temperature and time (°C)	Size of particles (μm)	BET surface area (m^2/g)	Surface elements			
				C (%)	O (%)	Si (%)	P (%)
AC30	650 °C/30 min	17.4889	25.2337	72.45	23.91	3.42	0.22
AC45	650 °C/45 min	23.3497	44.6221	74.20	22.52	3.09	0.19
AC60	650 °C/60 min	18.7503	108.9202	80.18	17.82	1.04	0.96

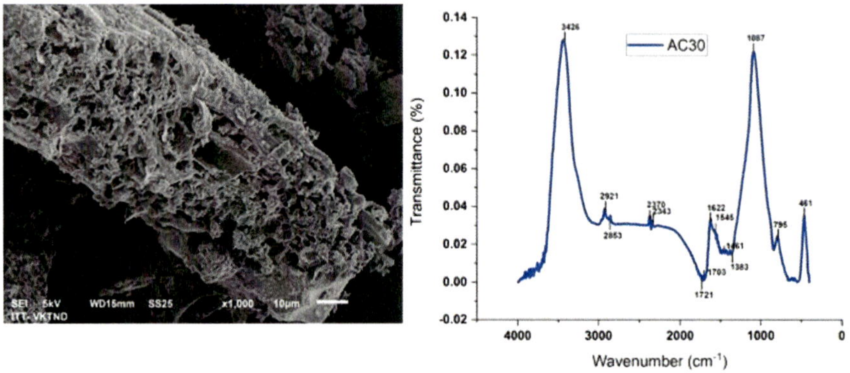

Fig. 3 SEM images and FTIR spectra of AC30 material activated from bamboo leaves at 650 °C/30 min

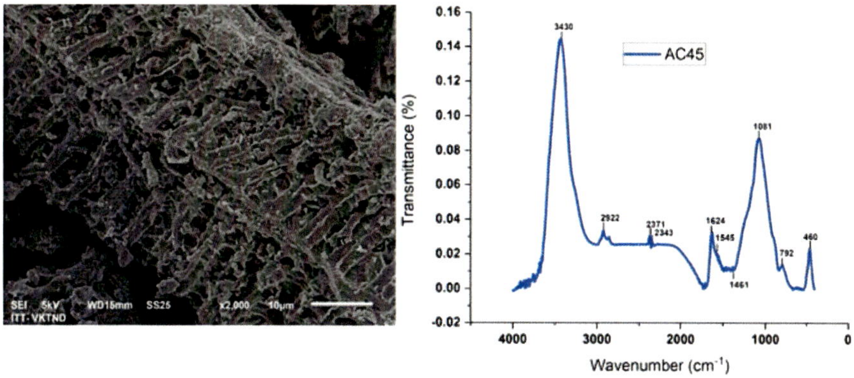

Fig. 4 SEM images and FTIR spectra of AC45 material activated from bamboo leaves at 650 °C/45 min

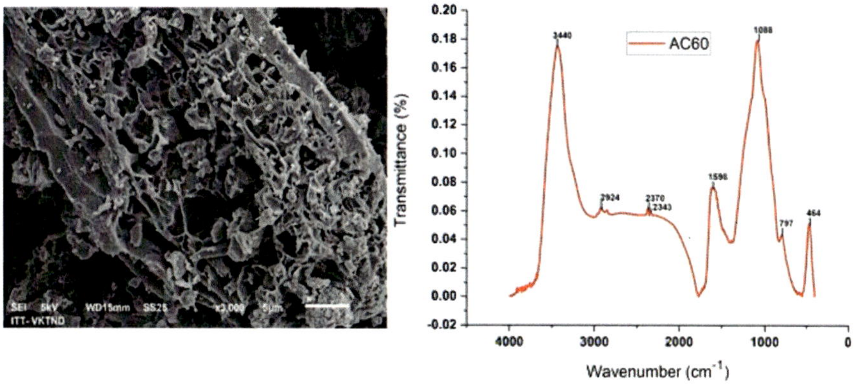

Fig. 5 SEM images and FTIR spectra of AC60 material activated from bamboo leaves at 650 °C/60 min

followed by the AC45 and AC30 samples with 74.2 and 72.45% C, respectively. The C content plays an important role in creating small porous with large volume to increase the adsorption surface area for activated carbon material [41]. The elemental percentage composition in Table 2 also showed that three activated carbon material AC30, AC45, AC60 contained other elements such as: O were found to be with the lowest percentage from 17.82% (AC60 sample) to the highest of 23.91% (AC30 sample); element Si and P have lower percentages from 1.04 to 3.42% (Si) and 0.19 to 0.22% (P) respectively. All three samples have a relatively high BET surface area, special the AC60 sample has a BET value of up to 108.9202 m^2/g. With the average particle size ranging from 17.4889 μm (AC30) to 23.3497 μm (AC45) showed that the material is smooth particles and quite similar together, the particle size of the AC60 sample is more uniform [11]. The small particle size combined with the large specific surface area will affect the adsorption and removal of the material's pollutants [19].

The surface morphology of three activated carbon was studied by SEM technique, and the results showed that the surface structure of the material samples somewhat different and parallels with BET measurement results. It can be seen that the material structure in three samples is heterogeneous porous. However, the AC60 sample was observed to be most homogeneous, and it had a larger porous structure, more porous and contained the most significant surface area among the three samples. There was a significant positive between the current study and previous results [12, 47, 49]. It can be clearly observed the microporous with a larger pore diameter will lead to reduction in surface area and an increase in porous structure [49], improve adsorption capacity for azo dye. Compared to the raw materials, the activated carbon material has a more strongly adsorption capacity through a rough surface with tiny channels and porous structure [12, 29]. Similarly, the FTIR measurement results in Figs. 3, 4 and 5 indicated that the activation time and temperature influenced the material characteristics and the formation of functional groups on the material surface [37, 40]. All three samples have adsorption spectrum from 400 to 4000 cm^{-1} with characteristic bonds such as C=O, O–H, C–C, C=C... Apart from some functional groups, characteristic bonds lost or newly formed depending on activation time from 30, 45 or 60 min, the intensity of the characteristic peaks of samples also changed. The FTIR spectrum obtained for all three samples displayed the following bonds such as O–H stretch groups (peak at 3440–3426 cm^{-1}) and C–N amine (peak at 1088–1881 cm^{-1}). The absorption peaks at 2924–2853 cm^{-1} and 1680–1620 cm^{-1} are attributed to the C–H stretch and C=C groups, respectively. The FTIR results also show peaks varied from 797 to 792 cm^{-1} for aromatics out of plane bend C–H. The C–halogen (Br/I) bonds appeared with a peak at 464–460 cm^{-1} or N=O bonds representing spectral range from 1550 to 1300 cm^{-1}. The C–O–C group was also presented with a peak at 1120 to 850 cm^{-1}. However, the C=O bond as ketone group (peak at 1725–1705 cm^{-1}) was only observed in the AC30 sample. The AC30 sample had lower purity than the two remain samples and containing many functional groups on the material surface, leading to dispersed and reduced treatment efficiency. Moreover, according to Pavia et al. [40] FTIR spectrum of three activated carbon samples in this study is different from many other samples because of containing bonds around 2350 cm^{-1} representing the CO_2 group. Our results corroborate with published studies [11, 41]. According to Chan et al. [11], the C=C, O–H, C–O–C groups play an important role in characteristic dye adsorption were represented for peaks at 1650, 3450, 1120 cm^{-1}, respectively. Furthermore, the study of Pongener et al. [41] also showed that the surface structure and pore size of the material are highly dependent on additional precursors during activation processing, in which the presence of H^+ ions will increase particle distribution, leading to an increase in porous structure and the surface area of the material.

3.2 Removal Efficiency of Azo Dye by Three Activated Carbon Materials

3.2.1 Brilliant Green

Effect of Initial Azo Dye Volume and Reaction Time on Adsorption Capacity

To study the effect of initial azo dye volume and reaction time on adsorption capacity, 0.01 g Brilliant green dye was diluted in 100 mL with distilled water and analyzed the initial concentration by HPLC instrument at wavelength 360 nm. The analyzed result recorded the actual concentrations of Brilliant green in the stock solution was 85.97 mg/L. The amount of Brilliant green dye adsorbed at the equilibrium by the three materials AC30, AC45 and AC60 is shown in Table 3. All samples recorded pH value changed from 9 to 9.4. When the dye volume added into the samples changed from 2, 3, 4 to 5 mL, and the reaction time increased from 30 to 120 min, the azo dye removal efficiency also increased accordingly. At the reaction time of 30 min, the adsorption capacity (q_e) of the AC30 sample was the lowest with 95.40 mg/g and the highest value was observed for the AC60 sample with 293.71 mg/g when added 2 and 5 mL Brilliant green, respectively. The adsorption capacity of Brilliant green azo dye reached a maximum value of 343.88 mg/g for the AC60 sample at reaction time 120 m with every volume ratio (2, 3, 4 and 5 mL). Similarly, the AC45 sample also recorded the maximum adsorption capacity (343.88 mg/L) when adding 2 and 3 mL Brilliant green at the reaction time 120 min, but when the volume increases to 4 and 5 mL, the efficiency decreased to 327.8 and 320.92 mg/g, respectively. With different reaction times (from 60 to 90 min), the complete removal capacity of azo

Table 3 The adsorption capacity q_e (mg/g) of Brilliant green dye by 3 activated carbon materials

Type of materials	$V_{azo\ dye}$ (mL)	$m_{activated\ carbon}$ (g)	q_e (mg/g)			
			Reaction time (minute)			
			30 min	60 min	90 min	120 min
AC30	2	0.5	237.95	265.28	297.00	343.88
AC45	2	0.5	255.57	299.86	333.93	343.88
AC60	2	0.5	293.71	337.37	343.88	343.88
AC30	3	0.5	186.11	201.89	257.95	326.27
AC45	3	0.5	201.92	286.47	320.92	343.88
AC60	3	0.5	285.96	328.04	343.88	343.88
AC30	4	0.5	166.40	197.95	237.96	309.43
AC45	4	0.5	186.11	265.00	299.85	327.80
AC60	4	0.5	265.00	324.74	343.88	343.88
AC30	5	0.5	95.40	124.00	151.16	297.78
AC45	5	0.5	146.67	229.78	265.00	320.92
AC60	5	0.5	257.62	297.43	320.91	337.37

dye was found for the AC60 sample with 2 mL Brilliant green adding and reaction time 60 min. Unlike the results for the AC60 sample, the two samples were AC30 và AC45 had reported the lowest adsorption capacity of 124 and 333.93 mg/g when added 5 mL (AC30 sample, reaction time 60 min) and 2 mL azo dye (AC45 sample, reaction time 90 min), respectively. It can be clearly seen that the volume ratio of Brilliant green: distilled water (mL) of 2:18 recorded the maximum removal capacity because the adsorption capacity has obtained 237.95 mg/g after only 30 min reaction time and reached the maximum capacity of 343.88 mg/g at all remaining reaction times (60, 90 and 120 min).

Effect of Adsorbent Material Mass and Reaction Time on Adsorption Capacity

Based on the result obtained in Sect. 3.2.1, the volume ratio of brilliant green: distilled water (mL) of 2:18 selected for further experiments. Figure 6 showed that when increasing the amount of the adsorbent from 0.25; 0.5; 1 to 2 g, the adsorption efficiency (H%) for azo dye increased as well. Only the adsorption efficiency obtained from the AC60 sample after 30 min reached 97.4% (with 0.25 g adsorbent material) and up to 100% as soon as this amount was increased to 0.5 g, while the remaining

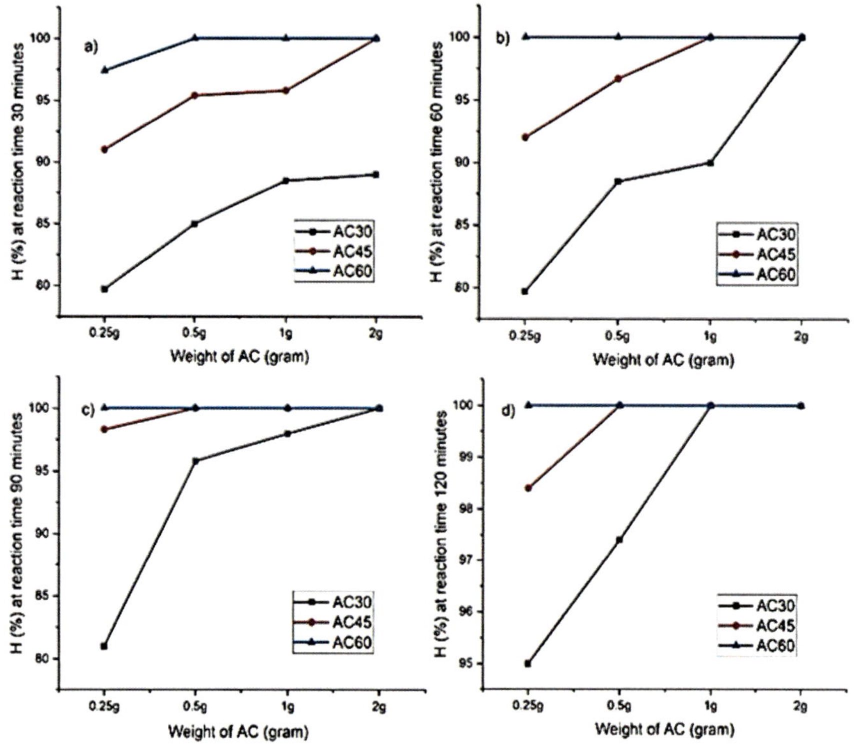

Fig. 6 The adsorption efficiency (H%) of Brilliant green dye from aqueous solution when adsorbed with 0.25, 0.5, 1 and 2 g activated carbon material

two samples AC30 and AC45 removed only 79.7 and 100% Brilliant green when added 0.25 and 2 g respectively, and even the AC30 sample achieved only 89% with the amount of adsorbent material 2 g. When the reaction time varied from 60 to 120 min and with just 0.25 g of adsorption material, the AC60 is the only sample that recorded the maximum adsorption efficiency of 100%, and the two remaining samples obtained the values of 79.7; 81; 95% (AC30 sample) and 94; 98.3; 98.4% (AC45 sample) with the reaction time 60, 90 and 120, respectively. Unlike the results for 0.25 g absorbent material, the adsorption efficiency with 0.5, 1 and 2 g for three activated carbon samples is very different together. The azo dye removal efficiency of the AC45 sample was recorded value up to 100% just as 0.5 g at reaction time 60 min, whereas the AC30 sample reached only the value of 100% with 2 g at 60 min. With the reaction time of 120 min and the amount of absorbent material 2 g, all three materials observed to able to remove 100% of Brilliant green azo dye from water. Of these, the AC60 is the only sample that obtained the completely removal efficiency of 100% with 0.5 g absorbent material at the reaction time of 30 min.

3.2.2 Alizarin Red S

Effect of Initial Azo Dye Volume and Reaction Time on Adsorption Capacity

Similarly, the initial concentration of Alizarin Red S solution analyzed by HPLC at wavelength 260 nm was 98.6 mg/L. The pH value in every sample measured around 8.9 to 9.2. The Alizarin Red S azo dye removal capacity at different time periods was lower than Brilliant green although the maximum adsorption capacity value of the material is higher. The results in Table 4 showed that when the 2 mL of Alizarin Red S azo dye was added, the removal capacity of all three materials recorded the rather high value and reached the maximum adsorption capacity of 394.4 mg/g at 60, 90 and 120 min by AC360, AC45 and AC30 samples, respectively. However, when the azo dye volume was altered from 3, 4 to 5 mL, the removal capacity of all samples was significantly reduced. As seen for the AC60 sample, the removal capacity was the highest for 5 mL of 118.32 mg/g and the lowest for 3 mL of 78.88 mg/g, respectively; with 4 mL Alizarin Red S, the adsorption capacity was only reached 177.48 mg/g (at 30 min) to 216.92 mg/g (at 120 min). The remaining samples showed the removal capacity for Alizarin Red S really reduced. As with 30 min reaction time and the volume of Alizarin Red S ranged from 3, 4 to 5 mL, the AC30 and AC45 samples only reached 134.10; 70.99; 7.89 mg/g and 248.47; 118.32; 27.61 mg/g, respectively. When the reaction time increased to 120 min, the adsorption capacity increased as well but only reached 39.44 to 256.36 mg/g (AC30 sample) and 78.88 to 343.13 mg/g (AC45 sample), respectively. It is also clear that the azo dye volume needed to reach the highest adsorption capacity value was 2 mL Alizarin Red S. This volume ratio would be selected for further experiments.

Table 4 The adsorption capacity q_e (mg/g) of Alizarin red S dye by activated carbon material

Type of materials	$V_{azo\ dye}$ (mL)	$m_{activated\ carbon}$ (g)	q_e (mg/g)			
			Reaction time (minute)			
			30 min	60 min	90 min	120 min
AC30	2	0.5	262.96	302.80	367.28	394.40
AC45	2	0.5	311.00	377.40	394.40	394.40
AC60	2	0.5	389.55	394.40	394.40	394.40
AC30	3	0.5	134.10	181.42	248.47	256.36
AC45	3	0.5	248.47	339.18	343.13	343.13
AC60	3	0.5	291.86	339.18	351.02	354.96
AC30	4	0.5	70.99	78.88	106.49	118.32
AC45	4	0.5	118.32	138.04	145.93	157.76
AC60	4	0.5	177.48	185.37	197.20	216.92
AC30	5	0.5	7.89	19.72	39.44	39.44
AC45	5	0.5	27.61	39.44	39.44	78.88
AC60	5	0.5	78.88	78.88	98.60	118.32

Effect of Adsorbent Material Mass and Reaction Time on Adsorption Capacity

As with Brilliant green azo dye, the adsorption efficiency for Alizarin Red S increases when increasing the reaction time and adsorbent material. The result of the experiments in Fig. 7 showed that the effect of adsorbent material mass and reaction time on adsorption capacity was lower than Brilliant green azo dye. With the volume ratio of Alizarin Red S: distilled water is 2:18 and the amount of the activated carbon material varied from 0.25 to 2 g, the adsorption efficiency of the material samples increases accordingly. When 0.25 g adsorption material was added, the AC30 sample recorded the lowest efficiency of 2%, and the highest was the AC60 with 30% at reaction time of 30 and 120 min, respectively. At a reaction time 30 min and the amount of adsorbent material increases from 0.5 to 1 g, the AC60 sample recorded the highest azo dye removal efficiency 74% (with 1 g adsorbent material) and the lowest value of 45% (with 0.5 g adsorbent material), the remaining two samples have the lowest removal capacity of 18% (with 0.5 g AC30 material sample) and the highest is 63% (with 1 g AC45 material sample). Between 60 and 120 m, the removal rate increases steadily but quite slowly. The maximum efficiency reached to 30, 45, 55% and 65, 87, 90%, respectively when added 0.5 and 1 g adsorbent material AC30, AC45, AC60 at reaction time 120 min. However, when the adsorbent material increased up to 2 g, the adsorption efficiency of all three samples was achieved to the maximum value, completely removing 100% of azo dye at all four reaction times (from 30 to 120 min) except for the AC30 sample with only 98% at 30 min.

In this way, it can be seen that the removal efficiency for Brilliant green azo dye better than Alizarin red S. The results also showed that the optimal experimental condition for the azo dye volume ratio/distilled water of 2/18 (mL). The adsorption

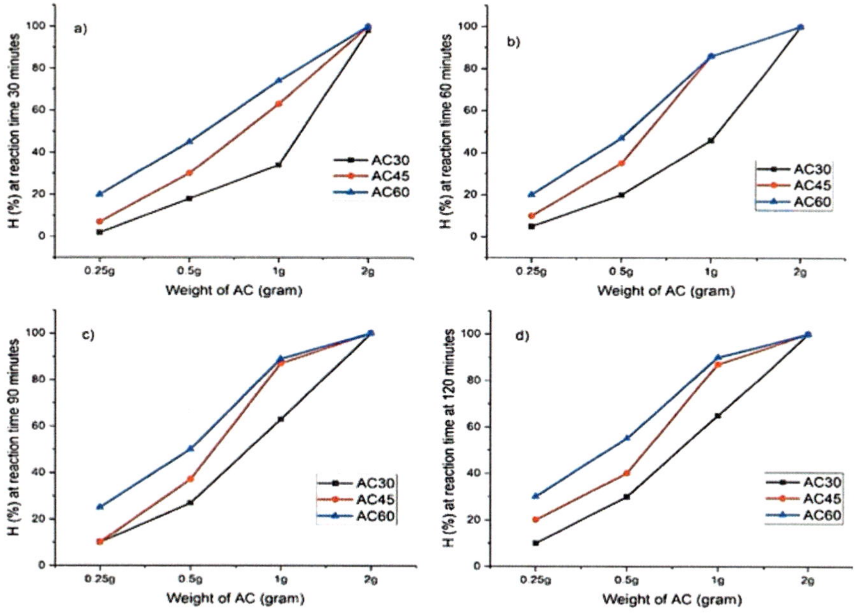

Fig. 7 The adsorption efficiency (H%) of Alizarin Red S dye from aqueous solution when adsorbed with 0.25, 0.5, 1 and 2 g activated carbon material

efficiency increases linearly with the reaction time and the weight of the adsorbent material added. However, only the AC60 sample achieved the maximum efficiency of 100% with both two azo dyes at reaction time of 30 min when the 0.5 and 2 g absorbent material for Brilliant green and Alizarin red S were added to the solution, respectively. This experimental result matches with the material structure characteristic of the AC60 sample. Further tests carried out with the optimization experimental conditions confirmed with our initial findings.

Azo dye is a group of dyes represented by a functional group (–N=N–) and the removal of this group from textile dyeing wastewater is a very difficult task. Our results have a number of similarities with some previously published findings [11, 19, 22]. Kaya and Uzun [22] used the carbonaceous material consist of pine cone, walnut shell, and hazelnut shell that were synthesized at different temperature range from 400 to 700 °C for 60 min to remove Alizarin yellow GG azo dye from the aqueous solution. When 20 ppm Alizarin red S and 8 g/L adsorbent material added to the solution at pH 3, the activated carbon sample from the walnut shells with the largest material surface area of 259.74 m^2/g had recorded the maximum adsorption efficiency of 82%. Similarly, Chan et al. [11] indicated that the dyes with a smaller molecular size would more easily adsorb onto the carbonaceous material and the dye removal efficiency depends on the appearance of the adsorbent functional groups on the material surface. In his investigation into the removal percentage of two azo dyes, Acid Yellow 117 (AY117) and Acid Blue25 (AB25), Chan et al. (2006) shows that the AB25 dye with smaller molecular size readily adsorbed onto the carbonaceous material than AY117 dye with larger size showed minimal adsorption. The results of material characterization also showed that the FTIR spectrum of activated carbon samples appeared dye adsorption functional groups such as C=C, O–H, C–O–C represented for peaks at 1650, 3450, 1120 cm^{-1}, respectively. The surface structure and pore size of the material play an important role in the adsorption and removal of the dye. Hameed and El-Khaiary [19] demonstrated that activated carbon samples have a high BET specific surface area, large pore size, the adsorption capacity with pollutants also increases accordingly. In his analysis, Hameed and El-Khaiary [19] point out the adsorption speed of malachite green (MG) is controlled by the pore structure of the material and the maximum removal efficiency reaches 263.58 mg/g with 300 mg/L of the absorbent material (synthesized from bamboo leaves at 850 °C/2 h) at reaction time 230 min.

For batch adsorption, the reaction time and the concentration of the initial adsorbent are important factors in determining the adsorption capacity value (q_e). When the adsorption time increases, the adsorption efficiency also increases accordingly [27, 43]. These fit well with the statement of Saeed et al. [27], Laskar and Kumar [43] and also confirm our findings. When the amount of the initial adsorption time increases from 0.25 to 2 g, the AC30 sample also recorded maximum efficiency, completely removing 100% azo dye after only 30 min of reaction time (BG dye) and 60 min (AR dye), the AC60 and AC45 samples also reached this value with 0.25 and 2 g, respectively, after only 30 min of reaction time. As mentioned by Laskar and Kumar [27], the removal efficiency of Brilliant green dye by the activated carbon material synthesized from Bambusa Tulda bamboo up to 98% only with 10 g/L of the absorbent material at the reaction time of 60 min. The Brilliant green removal percentage has been reduced to 83.74% when the amount of adsorbent material is reduced to 50 mg/L accordingly. Furthermore, the different activated carbon material types exhibit different adsorption efficiency. Laskar and Kumar [27] also demonstrated that the Brilliant green azo dye removal rate reduced from 136 mg/L (absorbed with the material modified with sodium carbonate) to 72 mg/L (the material modified with hydrochloric acid). Our findings appear to be well supported by Saeed et al. [43]. He underlines that the sorption equilibrium of crystal violet dye (CV) by the activated carbon from grapefruit peel reached rapidly to 96% after 60 min. The maximum adsorption capacity increases gradually from 60.42 to 254.16 mg/g as the adsorbent material content increases from 25 mg/L to 1 g/L, and the amount of CV added to the solution increases correspondingly from 10 to 600 mg/L.

As with changes in the reaction time, the solid/solution ratio also is an important factor in calculating the capacity of adsorption of azo dye in a batch treatment system. This confirms previous findings in the literature of Ahmad et al. [3], Al-Da'amy and Al-Shemary [4], Xu et al. [50]. According to Xu et al. [50], the methyl violet (MV) adsorption of carbonaceous materials synthesized from 156 g rice husks achieved the highest efficiency with the ratio of MV/distilled water of 1/18.2 (mmol/L). Xu et al. [50] demonstrated that the adsorption of methyl violet by the carbonaceous material is related to electrostatic attraction, surface precipitation, the specific interactions between dyes and hydroxyl, and carboxylate groups on the surface material.

Ahmad et al. [3] also showed that methylene blue (MB) was completely removed by the carbonaceous material prepared from rice husk, cow dung and sludge. When the 0.5, 1.0, 1.5, 3.0, and 6.0 g of the absorbent material added to 100 mL solution containing 100 mg/L of MB, the maximum efficiency achieved with three materials: rice husk, cow dung and sludge of 97.0–99.0, 71.0–99.0 and 73.0–98.9%, respectively. Similarly, the removal capacity of Alizarin yellow R and Alizarin red S by the absorbent material from snail shells recorded the maximum value after 30 min with the amount of initial absorbent material of 0.01 g and the initial dye concentration of 30 mg/L [4].

The azo dye removal efficiency of the carbonaceous materials also depends on the adsorption mechanism. Mukarram et al. 2020 reported on adsorption for methyl orange (MO) and Eriochrome Black-T (EBT)—and cationic dyes—methylene blue (MB) and crystal violet (CV) and concluded that the maximum adsorption capacity reached up to 206.61, 309.59, 163.132 and 934.57 mg/g when supplemented with 200 mg/L MB, EBT, MO, and CV, respectively. According to Mukaram et al. (2020), the adsorption mechanism is mainly based on the Redlich-Peterson isotherm model ($R^2 > 0.95$). However, Saniyaa et al. [44] used the isothermal model Langmuir and Freundlich to demonstrate the optimal experimental parameters to remove crystal violet (CV) dye from the textile dyeing wastewater. Saniyaa et al. [44] showed that the maximum removal efficiency of 70% had obtained after reaction time of 60 min with 100 mg/L of the carbonaceous material synthesized from curry bark and the initial dye concentration of 50 mg/L. He comes to reach the conclusion that the Langmuir isothermal model is most suitable in his work. Three kinetic models,namely, pseudo-first-order, pseudo-second-order, and intra-particle diffusion were used to analyze the adsorption mechanism. The kinetic results showed that the pseudo-second-order equation was the best model. The isotherm analysis indicated that the equilibrium data were well fitted to the Langmuir isotherm model, showing a monolayer adsorption manner of the dyes onto a homogeneous surface of the modified nanoparticles. According to the experimental results, about 97.8% of alizarin yellow and 78.7% of alizarin red were removed from aqueous solutions under optimal conditions [18].

3.3 *Effect to Actual Textile Dyeing Wastewater Sample and Compare the Material Quality with Some Commercial Activated Carbon Samples*

3.3.1 Effect on Initial Textile Dyeing Wastewater Sample

To evaluate the application capacity of the activated carbon sample, an initial wastewater sample (NT0) was collected from a factory of jeans in Long Bien district, Hanoi city, Vietnam and moved to the laboratory. Initial wastewater samples were measured for pH and analyzed the concentration of azo dye in the solution. The experimental

Table 5 The analysis results with initial wastewater samples at different time periods

No	Items	NT0	NT30	NT60	NT90	NT120
1	pH	8.4	9.2	9.0	9.0	9.0
2	BG	66.29 mg/L	0	0	0	0
3	AR	0 mg/L	0	0	0	0

optimization conditions selected in Sect. 3.2 will be applied with the initial wastewater samples at different time range from 30 min (NT30), 60 min (NT60), 90 min (NT90) to 120 min (NT120) and 0.5 g the absorbent material AC60. The analysis results are shown in Table 5 and Fig. 8.

The analysis result showed that the Brilliant green dye was completely removed from the solution after only 30 min of reaction time, HPLC spectrum did not record the appearance of a peak at the retention time of 16.232 min for samples absorbed at 60, 90, and 120 min, respectively. As reported above, this is a factory of jeans so the analysis result was no peak of AR dye. Both samples before and after absorbed by AC60 material did not record a peak at the retention time of 3.364 min.

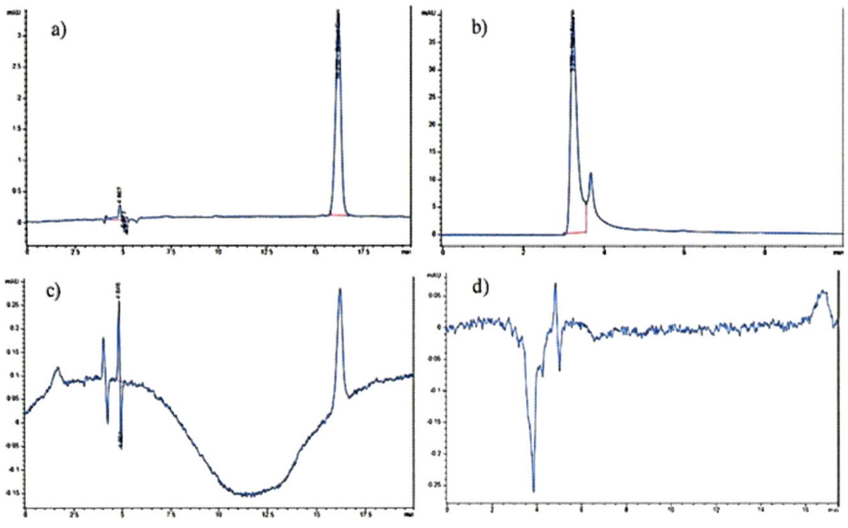

Fig. 8 HPLC (High Performance Liquid Chromatography) spectrum of standard and initial sample: **a** standard spectrum with Brilliant green dye; **b** standard spectrum with Alizarin red S dye; **c** spectrum of wastewater sample before adsorption to AC60 material and **d** spectrum of wastewater sample after adsorption with 0.5 g of AC60 material at 30 min

3.3.2 Compare the Material Quality with Some Commercial Activated Carbon Samples

To evaluate the quality of material in the current study, the AC60 sample with the best treatment efficiency would be compared with four other commercial material types available in the Vietnam market consists of AC-R (Russian activated carbon), AC-C (Chinese activated carbon), AC-F (French activated carbon) and AC-Tra Bac (Vietnamese activated carbon from coconut shell). The adsorption efficiency of four materials for Brilliant green and Alizarin red was performed according to the experimental optimization condition in Sect. 3.2. The results are shown in Table 6 and Fig. 9 as follows:

From the comparison results, it can be seen that the adsorption efficiency for BG and AR dyes by four commercial activated carbon materials share a number of similarities with our findings with AC60 material in the same optimization experimental conditions. The analysis results also showed that the azo dye removal efficiency of the AC60 material synthesized from bamboo leaves is higher than some

Table 6 The adsorption capacity (q_e) for BG and AR by four commercial activated carbon material compared to AC60 material

No	Type of materials	$V_{azo\ dye}$ (mL)	$m_{activated\ carbon}$ (g)	pH	q_e (mg/g)	
					Brilliant green (at 30 min)	Alizarin red S (at 60 min)
1	AC-C	2	0.5	4	311.20	325.60
2	AC-R	2	0.5	4	343.88	394.40
3	AC-F	2	0.5	3	343.88	394.40
4	AC-Tra Bac	2	0.5	5	343.88	394.40
5	AC60	2	0.5	9.1	343.88	394.40

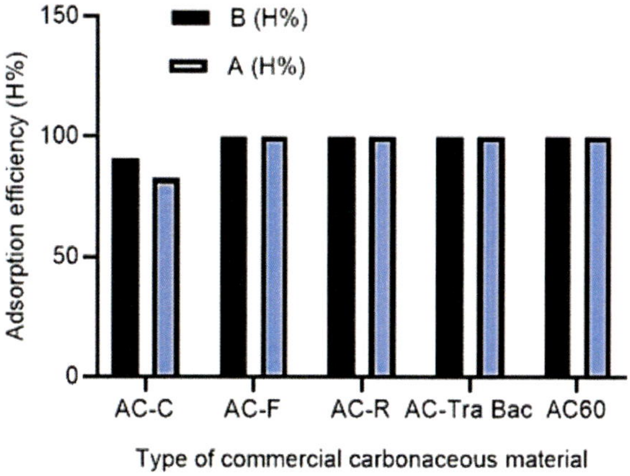

Fig. 9 The adsorption efficiency of four commercial activated carbon materials compared to AC60 material

previously prepared material (AC-C material reached only 90.5 and 82.55% for Brilliant green and Alizarin red S, respectively). As anticipated, our experiments prove that the removal efficiency for azo dye depends on experimental conditions such as reaction time, pyrolysis temperature, pH … and the activated carbon material quality. According to Mahdi et al. [33], the pyrolysis time will determine the material recovery productivity, affect the structure of cellulose and hemicellulose and release of organic molecules. The formation of functional groups (C=O, C–O–C and C–O) will determine the azo dye removal efficiency and therefore improve the material's surface characteristics (surface area and pore volume). The removal efficiency of pollutants depends on the functional groups on the carbonaceous material surface such as hydroxyl, carboxyl and –OH, C–H, C=O… These functional groups will have a strong influence on ion adsorption capacity [21].

The material synthesis environment and the experimental conditions were greatly affected the maximum adsorption capacity [31]. The pH value in the experimental medium is in contradiction with previous results reported in the documents about its effect on pollutant removal efficiency. Gautam et al. [17] suggested that the acidic environment enhances sorption processing. The maximum adsorption capacity of material from different waste plant biomass (consists of orange peel, flower waste, and environmentally harmful Alligator weed) recorded at 50 °C were 42.58 and 68.78 mg/g for Alizarin red S and Tartrazine at pH 4.0 and 2.0, respectively. Our findings do not support the data provided by Gautam et al. [17] in this area. In fact, in contrast with what was previously data, we found that the pH value in the current study change around pH 9 and our findings appear to be well supported by Kyi's report [26]. Kyi et al. [26] carried out the sorption experiment to remove crystal violet from wastewater with adsorbent material from palm kernel shell. The results showed that the removal percentage and adsorption capacity of the material increases linearly with the pH value in the solution. These differences can be explained partly by the negative charges on the surface of carbonaceous materials at high pH enhance electrostatic attraction between azo dye molecules and carbonaceous material.

According to Sumalinog et al. [46], the removal efficiency of acetaminophen (APAP) and methylene blue (MB) dyes from aqueous solution were affected by changes in pH value. The removal efficiency of MB dye (99.9%) recorded the higher value than APAP dye (only reached 63.7%). While the adsorption capacity of APAP was strongly reduced by pH change, MB adsorption found to be unaffected when pH increases from 2 to 12. Chahinez et al. [10] argued that the concentration of crystal violet dye (CV) adsorbed into the carbonaceous material derived from date palm petioles increased by 46% (from 18.8 to 27.4 mg/g, an average of 24.36 mg/g) when the solution pH increased from 2.0 to 12. Zazycki et al. [53] showed that the treatment capacity of methyl violet dye (MV) is higher in alkaline medium. Surprisingly, for maximum adsorption capacity of the carbonaceous material from chitin was found up to 1120.8 mg/g with pH changed from 6.8 to 9. Consequently, it can be seen that the characteristics and source of the raw materials have a significant influence on the carbonaceous material structure, determining the treatment efficiency as well as the removal capacity for pollutants. Our study provides additional further evidence for insight into the material synthesized conditions such as reaction time, pyrolysis temperature as well as the volume ratio of dye/water added to the solution for azo dye removal in textile wastewater.

4 Conclusion

This study investigated the removal capacity of two azo dyes, including Brilliant green and Alizarin red S, by three types of activated carbon prepared from bamboo leaves, AC30 (650 °C/30 min); AC45 (650 °C/45 min) and AC60 (650 °C/60 min). The obtained materials have relatively high C content (AC30 is 72.45%; AC45 is 74.30% and AC60 is 80.18%) and contain the characteristic bonds represented by the azo dye removal capacity C=C; C–O–C, –OH. The results also showed that the AC60 material sample has the best adsorption efficiency reached 100% at the reaction time of 30 min with the volume ratio of azo dye/distilled water of 2/18 (mL), pH 9, the absorbent amount of 0.5 and 2 g for Brilliant green and Alizarin red S, respectively.

The analysis results of initial textile dyeing wastewater sample with four commercial activated carbon samples showed that the material source and the experimental conditions in the current study are suitable for the azo dye removal application from industrial textile dyeing wastewaters. Despite this, we believe our work could be a starting point for further studies on the adsorption mechanism, the wastewater properties, desorption/adsorption… need to be thoroughly conducted before application in factories or actual conditions.

References

1. Abd-Elhamid AI, Emran M, El-Sadek MH, El-Shanshory AA, Soliman HMA, Akl MA, Rashad M (2020) Enhanced removal of cationic dye by eco-friendly activated biochar derived from rice straw. Appl Water Sci 10(1):45. https://doi.org/10.1007/s13201-019-1128-0
2. Aboel-Zahab H, El-Khyat Z, Sidhom G, Awadallah R, Abdel-Al W, Mahdy K (1997) Physiological effects of some synthetic food colouring additives on rats. Boll Chim Farm 136(10):615–627
3. Ahmad A, Khan N, Giri BS, Chowdhary P, Chaturvedi P (2020) Removal of methylene blue dye using rice husk, cow dung and sludge biochar: characterization, application, and kinetic studies. Biores Technol 306:123202. https://doi.org/10.1016/j.biortech.2020.123202
4. Al-Da'amy MA, Al-Shemary RQ (2018) Removal of alizarin red dye from aqueous solution with bio sorption technique using snail shell as low cost adsorbent. J Glob Pharma Technol 422–430
5. Amirza MAR, Adib MMR, Hamdan R (2017) Application of agricultural wastes activated carbon for dye removal—an overview. In: MATEC web of conferences, vol 103, p 06013. https://doi.org/10.1051/matecconf/201710306013
6. Ashraf MW, Abulibdeh N, Salam A (2019) Adsorption studies of textile dye (chrysoidine) from aqueous solutions using activated sawdust. Int J Chem Eng 2019:1–8. https://doi.org/10.1155/2019/9728156
7. Bello OS, Adegoke KA, Olaniyan AA, Abdulazeez H (2015) Dye adsorption using biomass wastes and natural adsorbents: overview and future prospects. Water Treat 53:1292–1315. https://doi.org/10.1080/19443994.2013.862028

8. Benkhaya S, M'rabet S, El Harfi A (2020) Classifications, properties, recent synthesis and applications of azo dyes. Heliyon 6(1):e03271. https://doi.org/10.1016/j.heliyon.2020.e03271
9. Bhomick PC, Supong A, Baruah M, Pongener C, Gogoi C, Sinha D (2019) Alizarin Red S adsorption onto biomass-based activated carbon: optimization of adsorption process parameters using Taguchi experimental design. Int J Environ Sci Technol 17:1137–1148. https://doi.org/10.1007/s13762-019-02389-1
10. Chahinez HO, Abdelkader O, Leila Y, Tran HN (2020) One-stage preparation of palm petiole-derived biochar: characterization and application for adsorption of crystal violet dye in water. Environ Technol Innov 100872. https://doi.org/10.1016/j.eti.2020.100872
11. Chan LS, Cheung WH, McKay G (2008) Adsorption of acid dyes by bamboo derived activated carbon. Desalination 218(1–3):304–312. https://doi.org/10.1016/j.desal.2007.02.026
12. Cui J, Yang Y, Hu Y, Li F (2015) Rice husk based porous carbon loaded with silver nanoparticle by a simple and cost-effective approach and their antibacterial activity. J Colloid Interface Sci 455:117–124. https://doi.org/10.1016/j.jcis.2015.05.049
13. Dias YN, Souza ES, da Costa HSC, Melo LCA, Penido ES, do Amarante CB, Teixeira OMM, Fernandes AR (2019) Biochar produced from Amazonian agro-industrial wastes: properties and adsorbent potential of Cd^{2+} and Cu^{2+}. Biochar 1:389–400. https://doi.org/10.1007/s42773-019-00031-4
14. Edward LKM, Cheung WH, Valix M, McKay G (2010) Dye adsorption onto char from bamboo. J Hazard Mater 177(1–3):1001–1005. https://doi.org/10.1016/j.jhazmat.2010.01.018
15. El maguana Y, Elhadiri N, Benchanaa M, Chikri R (2020) Activated carbon for dyes removal: modeling and understanding the adsorption process. J Chem 2020:1–9.https://doi.org/10.1155/2020/2096834
16. Fei B, Gao Z, Wang J, Liu Z (2016) Biological, anatomical, and chemical characteristics of bamboo. Econ Appl Second Xylem 283–306. https://doi.org/10.1016/B978-0-12-802185-9.00014-0
17. Gautam PK, Shivapriya PM, Banerjee S, Sahoo AK, Samanta SK (2019) Biogenic fabrication of iron nanoadsorbents from mixed waste biomass (MWB) for aqueous phase removal of alizarin red S and tartrazine: kinetics, isotherm and thermodynamic investigation. Environ Progr Sustain Energy 1–13. https://doi.org/10.1002/ep13326
18. Gholivand MB, Yamini Y, Dayeni M, Seidi S, Tahmasebi E (2015) Adsorptive removal of alizarin red-S and alizarin yellow GG from aqueous solutions using polypyrrole-coated magnetic nanoparticles. J Environ Chem Eng 3:529–540. https://doi.org/10.1016/j.jece.2015.01.011
19. Hameed BH, El-Khaiary MI (2008) Equilibrium, kinetics and mechanism of malachite green adsorption on activated carbon prepared from bamboo by K_2CO_3 activation and subsequent gasification with CO_2. J Hazard Mater 157(2–3):344–351. https://doi.org/10.1016/j.jhazmat.2007.12.105
20. Hao Y, Wang Z, Gou J, Dong S (2015) Highly efficient adsorption and removal of Chrysoidine Y from aqueous solution by magnetic graphene oxide nanocomposite. Arab J Chem. https://doi.org/10.1016/j.arabjc.2015.07.013
21. Hirata M, Kawasaki N, Nakamura T, Matsumoto K, Kabayama M, Tamura T, Tanada S (2002) Adsorption of dyes onto carbonaceous materials produced from coffee grounds by microwave treatment. J Colloid Interface Sci 254:17–22. https://doi.org/10.1006/jcis.2002.8570
22. Kaya N, Uzun ZY (2020) Investigation of effectiveness of pyrolysis products on removal of alizarin yellow GG from aqueous solution: a comparative study with commercial activated carbon. Water Sci Technol 1191–1208. https://doi.org/10.2166/wst.2020.213
23. Khamis MI, Ibrahim TH, Jumean FH, Sara ZA, Atallah BA (2020) Cyclic sequential removal of alizarin Red S dye and Cr(VI) ions using wool as a low-cost adsorbent. Processes 8(5):556. https://doi.org/10.3390/pr8050556
24. Kornaros M, Lyberatos G (2006) Biological treatment of wastewaters from a dye manufacturing company using a trickling filter. J Hazard Mater 136(1):95–102. https://doi.org/10.1016/j.jhazmat.2005.11.018

25. Krishna Mohan GV, Naga Babu A, Kalpana K, Ravindhranath K (2016) Removal of naphthol green B dye from polluted waters using hydrogen peroxide treated red mud. Der Pharma Chem 8(19):403–414
26. Kyi PP, Quansah JO, Lee C-G, Moon J-K, Park S-J (2020) The removal of crystal violet from textile wastewater using palm kernel shell-derived biochar. Appl Sci 10(7):2251. https://doi.org/10.3390/app10072251
27. Laskar N, Kumar U (2018) Removal of Brilliant Green dye from water by modified *Bambusa Tulda*: adsorption isotherm, kinetics and thermodynamics study. Int J Environ Sci Technol 16:1649–1662. https://doi.org/10.1007/s13762-018-1760-5
28. Lee JW, Choi SP, Thiruvenkatachari R, Shim WG, Moon H (2006) Submerged microfiltration membrane coupled with alum coagulation/powdered activated carbon adsorption for complete decolorization of reactive dyes. Water Res 3:435–444. https://doi.org/10.1016/j.watres.2005.11.034
29. Liu N, Charrua AB, Weng C-H, Yuan X, Ding F (2015) Characterization of biochars derived from agriculture wastes and their adsorptive removal of atrazine from aqueous solution: a comparative study. Biores Technol 198:55–62. https://doi.org/10.1016/j.biortech.2015.08.129
30. Lobovikov M, Paudel S, Piazza M, Ren H, Wu J (2007) World bamboo resources. Food and Agriculture Organization of the United Nations 18(1): 70. https://doi.org/10.13140/RG.2.1.1042.3764
31. Macedo JdS, Júnior NBdC, Almeida LE, Vieira EFdS, Cestari AR, Gimenez IdF (2006) Kinetic and calorimetric study of the adsorption of dyes on mesoporous activated carbon prepared from coconut coir dust. J Colloid Interface Sci 298:515–522. https://doi.org/10.1016/j.jcis.2006.01.021
32. Mahamad MN, Zaini MAA, Zakaria ZA (2015) Preparation and characterization of activated carbon from pineapple waste biomass for dye removal. Int Biodeterior Biodegrad 102:274–280. https://doi.org/10.1016/j.ibiod.2015.03.009
33. Mahdi Z, El Hanandeh A, Yu Q (2017) Influence of pyrolysis conditions on surface characteristics and methylene blue adsorption of biochar derived from date seed biomass. Waste Biomass Valorization 8:2061–2073. https://doi.org/10.1007/s12649-016-9714-y
34. Michalina O, Ryan FD, Rita JM, Glenn K, Christopher TE (2011) The potential for human exposure, direct and indirect, to the suspected carcinogenic triphenylmethane dye Brilliant Green from green paper towels. Food Chem Toxicol 49:1870–1876. https://doi.org/10.1016/j.fct.2011.05.005
35. Mittal A, Mittal J, Malviya A, Gupta VK (2010) Removal and recovery of Chrysoidine Y from aqueous solutions by waste materials. J Colloid Interface Sci 344(2):497–507. https://doi.org/10.1016/j.jcis.2010.01.007
36. Nautiyal P, Subramanian KA, Dastidar MG (2016) Adsorptive removal of dye using biochar derived from residual algae after in-situ transesterification: alternate use of waste of biodiesel industry. J Environ Manag 182:187–197. https://doi.org/10.1016/j.jenvman.2016.07.063
37. Nguyen DT (2012) Modern physical methods applied in chemistry. National University Publishing House, Hanoi
38. Oveisi M, Asli MA, Mahmoodi NM (2018) MIL-Ti metal-organic frameworks (MOFs) nanomaterials as superior adsorbents: synthesis and ultrasound-aided dye adsorption from multicomponent wastewater systems. J Hazard Mater 347:123–140. https://doi.org/10.1016/j.jhazmat.2017.12.057
39. Pathania D, Sharma S, Singh P, Arabian J (2017) Removal of methylene blue by adsorption onto activated carbon developed from *Ficus carica* bast. Arab J Chem 10:S1445–S1451. https://doi.org/10.1016/j.arabjc.2013.04.021
40. Pavia DL, Lampman GM, Kriz GS, Vyvyan JR (2013) Introduction to spectroscopy. United States of America
41. Pongener C, Kibami D, Rao KS (2015) Sythesis and characterization of activated carbon from the biowaste of the plant manihot esculenta. Chem Sci Trans 59–68. https://doi.org/10.7598/cst2015.958

42. Rehman R, Mahmud T (2013) Sorptive elimination of alizarin Red-S dye from water using citrullus lanatus peels in environmentally benign way along with equilibrium data modeling. Asian J Chem 25(10):5351–5356. https://doi.org/10.14233/ajchem.2013.14179
43. Saeed A, Sharif M, Iqbal M (2010) Application potential of grapefruit peel as dye sorbent: kinetics, equilibrium and mechanism of crystal violet adsorption. J Hazard Mater 179(1–3):564–572. https://doi.org/10.1016/j.jhazmat.2010.03.041
44. Saniyaa A, Sathyaa K, Nagarajanb K, Yogesha M, Jayalakshmia H, Praveenaa P (2020) Modelling of the removal of crystal violet dye from textile effluent using Murraya koenigii stem biochar. Desalin Water Treat 203:356–365. https://doi.org/10.5004/dwt.2020.26191
45. Srinivasan A, Viraraghavan T (2010) Decolorization of dye wastewaters by biosorbents: a review. J Environ Manag 91(10):1915–1929. https://doi.org/10.1016/j.jenvman.2010.05.003
46. Sumalinog DAG, Capareda SC, de Luna MDG (2018) Evaluation of the effectiveness and mechanisms of acetaminophen and methylene blue dye adsorption on activated biochar derived from municipal solid wastes. J Environ Manag 210:255–262. https://doi.org/10.1016/j.jenvman.2018.01.010
47. Suresh S (2014) Treatment of textile dye containing effluents. Curr Environ Eng 1:1–23. https://doi.org/10.2174/2212717801666141021235246
48. Tran TTH, Nguyen VH, Vu NT, Nguyen XT, Tran AQ, Vu KT (2021) Initial results of using biochar derived from spent coffee grounds to remove pollutants from livestock wastewater in Vietnam. In book: Proceedings of the international conference on innovations for sustainable and responsible mining, vol 108, pp 305–325. https://doi.org/10.1007/978-3-030-60269-7_16
49. Vu TM, Trinh VT, Doan DP, Van HT, Nguyen TV, Vigneswaran S, Ngo HH (2017) Removing ammonium from water using modified corncob-biochar. Sci Total Environ 579:612–619. https://doi.org/10.1016/j.scitotenv.2016.11.050
50. Xu RK, Xiao SC, Yuan JH, Zhao AZ (2011) Adsorption of methyl violet from aqueous solutions by the biochars derived from crop residues. Biores Technol 102(22):22. https://doi.org/10.1016/j.biortech.2011.08.089
51. Yang G, Wu L, Xian Q, Shen F, Wu J, Zhang Y (2016) Removal of congo red and methylene blue from aqueous solutions by vermicompost-derived biochars. PLoS One 11(5):e0154562. https://doi.org/10.1371/journal.pone.0154562
52. Yang Y, Lin X, Wei B, Zhao Y, Wang J (2014) Evaluation of adsorption potential of bamboo biochar for metal-complex dye: equilibrium, kinetics and artificial neural network modeling. Int J Environ Sci Technol 11(4):1093–1100. https://doi.org/10.1007/s13762-013-0306-0
53. Zazycki MA, Borba PA, Silva RNF, Peres EC, Perondi D, Collazzo GC, Dotto GL (2019) Chitin derived biochar as an alternative adsorbent to treat colored effluents containing methyl violet dye. Adv Powder Technol 30(8):1494–1503. https://doi.org/10.1016/j.apt.2019.04.026
54. Zhu Y, Yi B, Yuan Q, Wu Y, Wang M, Yan S (2018) Removal of methylene blue from aqueous solution by cattle manure-derived low temperature biochar. RSC Adv 8(36):19917–19929. https://doi.org/10.1039/C8RA03018A

Removal of Rifampin by Luffa: A Pharmaceutical Potential in Producing Dye in Water

Mehrdad Negarestani, Amir Lashkari, Ali Khadir, and Afsaneh Mollahosseini

Abstract Pharmaceuticals are regarded as one of the prominent environmental concerns due to their entrance and detection in aquatic solutions. Some pharmaceuticals are able to generate colouring agents in water which must also be removed. This chapter focus on the removal of rifampin, a drug producing a reddish-orange to reddish-brown colour in water, by Luffa biomaterial. The bioadsorbent was characterized with SEM, XRD, and FTIR. In view of experimental tests, it turned out that equilibrium reached in 140 min, and at a concentration of 5 ppm, the removal efficiency of approximately 100% was achieved. The highest rifampin removal was observed at 25 °C. The kinetics and isotherm studies let out that the experimental results followed pseudo-second-order, Elovich, and Langmuir models. It is fair to suggest that based on the results, Luffa biomaterial could be nominated as a sustainable, non-toxic and rifampin removal that can adsorb it from aqueous solutions

Keywords Luffa · Rifampin · Biomaterial · Isotherm · Kinetics · Removal

1 Introduction

The word "Environmental Pollution" or especially "Water pollution" started with the entrance of pollutants such as heavy metals into different water media, posing

M. Negarestani
Department of Civil and Environmental Engineering, Iran University of Science and Technology, Tehran, Iran

A. Lashkari
Faculty of Civil, Water and Environmental Engineering, Power and Water University of Technology (Abbaspour University), Tehran, Iran

A. Khadir (✉)
Young Researcher and Elite Club, Yadegar-e-Imam Khomeini (RAH) Shahre Rey Branch, Islamic Azad University, Tehran, Iran

A. Mollahosseini
Research Laboratory of Spectroscopy & Micro and Nano Extraction, Department of Chemistry, Department of Chemistry, Iran University of Science and Technology, Tehran, Iran

© The Author(s), under exclusive license to Springer Nature Singapore Pte Ltd. 2021
S. S. Muthu and A. Khadir (eds.), *Novel Materials for Dye-containing Wastewater Treatment*, Sustainable Textiles: Production, Processing, Manufacturing & Chemistry,
https://doi.org/10.1007/978-981-16-2892-4_9

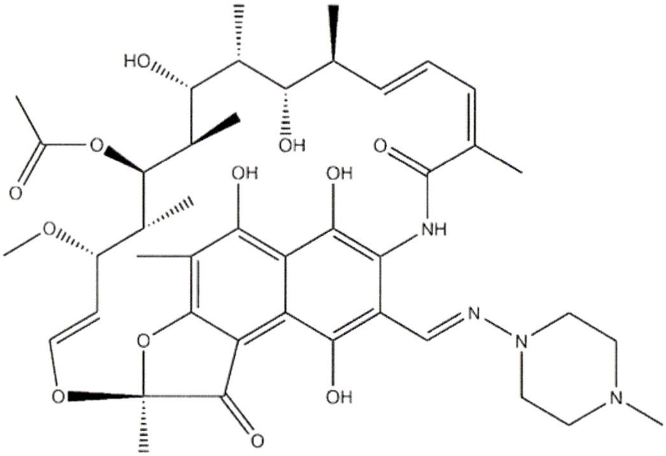

Fig. 1 Chemical structure of rifampicin

threats to the ecosystem. Currently, with the advancement in people's living standards, modernization, and agricultural activities, other new contaminants have been observed in waters, names emerging pollutants. Although emerging pollutants have so many categories and groups, pharmaceuticals and drugs are some of the most important ones. Aquatic-pharmaceutical pollution indicates the presence of drugs in waters.

Rifampicin or rifampin (Fig. 1), 3-[(4-methyl-1-piperazinyl)imino]-methyl, is a broad-spectrum antibiotic treating bacterial infection disease such as tuberculosis which requires 6–9 months' treatment period. It is also used for immunosuppressed patient's treatment. Since its introduction in 1968, it showed great capability in basic health care. Despite the fact that antibiotics have contributed to bringing long life for both humans and animals, they also have their own side effects that pose threats to the ecosystem [1]. With the rapid advancement in analytical techniques, many reports have declared that pharmaceuticals, including antibiotics, find their ways into our ecosystem [2, 3]. Excretion from humans, inappropriate pharmaceutical waste management, improper and over consumptions, and inability of current water and wastewater treatment plants are the main reasons associated with the observation of pharmaceuticals in waters. Considering pharmaceuticals and antibiotics proliferation in aquatic media as a serious global concern, many scholars in academia have attempted to propose an effective treatment technique for pharmaceutical removal [4–9]. Another important side effect of pharmaceuticals in waters is their colouring agents or colourant. Pharmaceutical companies employ colourant with pharmaceuticals with different aims, including better appearance and acceptability by patients, facile detection particularly by elderly, brand development and product identity [10, 11]. Rifampin, for instance, causes a reddish-orange to reddish-brown colour which aesthetically is not accepted and must be removed. In this direction, the elimination of some drugs is associated with the removal of the dye released.

The high solubility of rifampin along with its colour has gained the attention of many researchers to find suitable methods of removal. Henrique et al. [12] studied rifampicin antibiotic elimination by calcined *Mytella falcata* shells. The experimental data indicated a rapid adsorption process, finished in 30–45 min. The kinetics results followed pseudo-first-order model assumptions, and the ultimate removal efficiency of 95% was reported. In another report, magnetic Fe_3O_4 nanoparticles (111.8 m^2/g) were utilized in rifampicin removal and showed a removal percentage of 98.4% [13]. Lin et al. [14] investigated simultaneous elimination of rifampicin and Pb (II) by an economical and non-polluting material, Fe nanoparticles adsorbent [14]. The surface area of the adsorbent was found 37.3 m^2/g which provides suitable porous structures for the removal of pollutants. High removal efficiencies of 100% and 92% were reported for Pb (II) and rifampicin, respectively. Liu et al. [15] conducted an interesting research on rifampicin removal by rGO@nFe/Pd (reduced graphene oxide composited with bimetallic iron/palladium nanoparticles) nanocomposite. Once the reactor was equipped with Fenton oxidation, the removal efficiency increased from 79.9 to 85.7%. For actual wastewaters, rivers, aquaculture wastewater and domestic wastewater, these composites showed promising removal capability, 80.4, 77.9 and 70.2%, respectively. Other works, including Madivoli et al. [16], Xu et al. [17], and da Silva Duarte [18], were also focused on rifampin removal from waters. In spite of these studies, the authors think that there is still a gap in rifampin removal as a dye–producing compound, and there are other low–cost biomaterials that could also be effective in water purification.

The adsorption process is now addressed as an advanced sustainable water purification and wastewater treatment technique that has been widely studied during the last decade. Adsorption is a surface process that transports a molecule from fluid to a solid surface via physical forces and chemical bonds [5]. The adsorption process is the most common procedure in the wastewater and water treatment industry because of its advantages, such as cost-effectiveness, high efficiency, and simple operation. Since now, many scholars have suggested different materials as adsorbent, including activated carbon, graphene-based composites, carbon nitride, hydrogels, metal–organic framework, and manganese oxide nanoparticles. Regarding the high cost associated with specific adsorbent preparation/synthesis (such as activated carbon), biomaterials have gained high attention due to their availability, cheap, non-toxicity, and perfect performance [19]. The applicability of biomaterials for the elimination of heavy metals, dye, pharmaceuticals, and other less essential pollutants has already been demonstrated. Nonetheless, more research and exploration must be conducted to figure out the performance of biomaterials in interaction with pollutant molecules.

Luffa (Fig. 2) is a famous biomaterial that is known as a cheap, non-toxic, affordable, porous, and robust environmental pollutant adsorbent that can adsorb various types of adulterants such as dyes, pharmaceuticals, and heavy metals [19]. Luffa is a natural fiber that is commonly distributed on Earth, and it is usually famous as sponge gourd, bath and vegetable sponge, dishcloth gourd, and loofa. The most famous Luffa family member is Luffa cylindrica, and this type of Luffa is available in some countries like India, Brazil, Nepal, South Africa, Korea, and the USA. The washing and bathing industries used Luffa to produce some products such as mats,

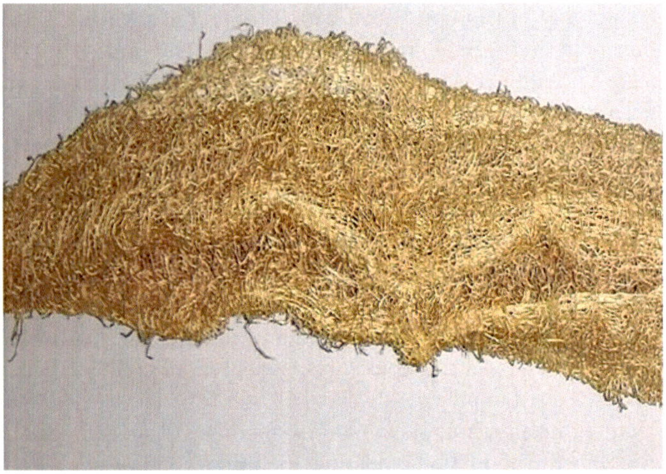

Fig. 2 Image of Luffa used in this study

bath towels, and sponge baths. Not only Luffa is used for washing purposes but also it is useful in the pharmaceutical industry. Luffa can adsorb heavy metals [20–22], pharmaceuticals [23, 24], dyes [25–27], etc. These works demonstrated that Luffa is a promising biomaterial in removing different kinds of pollutants.

In the present chapter, the authors employed Luffa biomaterial for the removal of dye produced from rifampin in aqueous solutions. In this direction, Luffa was initially characterized and then used in adsorption experiments. The kinetics and experimental data were analyzed by kinetics and isotherm models.

2 Luffa Preparation and Characterization

First, Luffa was bought from a cosmetic store located in Tehran, Iran. To extract impurities and dust from its top, it was washed a couple of times with purified water. Thereafter, it was inserted in an oven at 40 °C to become dry. Once it was fully dehydrated, Luffa was cut into similar pieces in order to be utilized in adsorption experiments.

SEM model TESCAN-Vega II was utilized for SEM images of Luffa biomaterial. On a SHIMADZU FT-IR 8400 S model, the spectrum of Fourier in frost transformation in Luffa biomaterial was reported. XRD model Philips PW 1800 Diffractomete was also employed for Luffa characterization.

3 Adsorption Experiments

In the current investigation, rifampin adsorption by Luffa adsorbent was tested. Rifampin stock solution (100 ppm) was formulated and stored in the refrigerator. A 50 mL Erlenmeyer flask was used to conduct the batch experiments. In order to optimize the process, the effects of contact time (0–170 min), initial rifampin concentration (5–50 ppm), Luffa dosage (0.1–05 g), and temperature (25–55 °C) were tested. Removal efficiency, adsorbed rifampin at equilibrium and time t were determined using the following equations:

$$Re\ (\%) = \frac{C_0 - C_t}{C_0} \times 100,$$

$$q_e = \frac{(C_0 - C_e) \times V}{m},$$

$$q_t = \frac{(C_0 - C_t) \times V}{m},$$

where $Re\ (\%)$ is the rifampin removal efficiency, C_0, C_t, and C_e are the initial, at time t, and equilibrium concentration of rifampin in the solution (ppm), respectively. V (L) is the volume of the working solution, and m (gr) is the mass of Luffa biomaterial used for rifampin removal.

4 Results and Discussion

4.1 Luffa Characterization

Characterization procedures are generally useful to study the features of biomaterials. In the present study, SEM images, FTIR spectrum, and XRD pattern are utilized to characterize Luffa biomaterial.

SEM images are useful images to see the surface texture of a material. Figure 3 shows SEM images of Luffa. Empty channels and holes are clearly observed on the surface of Luffa, providing empty sites for rifampin adsorption. The fibrous structure of Luffa could also be seen in these images which proves that this biomaterial belongs to fibrous adsorbents. These observations are in great conformity with previous studies, including Demir et al. [28] and El Ashtoukhy [29].

FTIR is an observational method that shows a range of absorption peaks used to analyze the existence of such functional groups. Extensive utilization of these techniques may attribute to the fact that it is rapid, economical, easy, and non-destructive [30]. The FTIR of Luffa is depicted in Fig. 4. The peak at 1626 cm^{-1} shows OH bending vibrations, and at 1399 cm^{-1}, C–O stretching was observed. Other peaks,

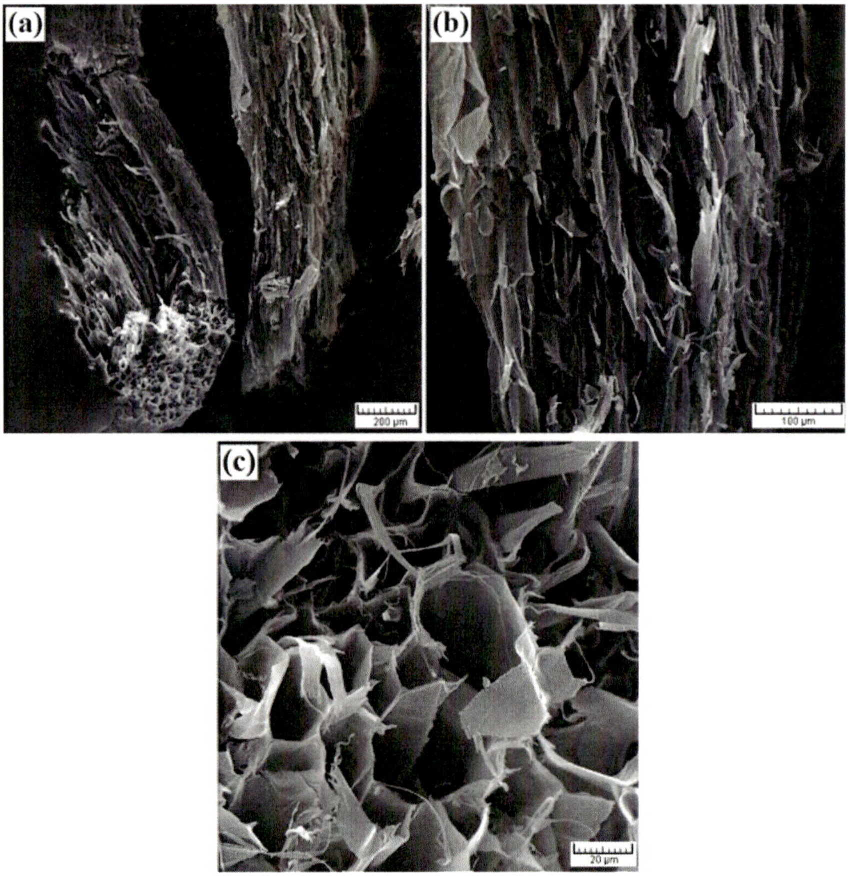

Fig. 3 SEM images of Luffa biomaterial used for rifampin removal

including 1089 and 1044 cm^{-1}, were indicative of cellulose. Similar peaks were reported in the previous literature [31, 32].

XRD is another nondestructive technique generally used for the identification of materials. XRD pattern of raw Luffa biomaterial is shown in Fig. 5. Important peaks are observed at 15.9° and 22.3°, and another peak at around 34.3° could be detected. In addition, at an angle lower than 10°, another peak is recorded. Zhang et al. [33] studied the reliability and efficiency of Luffa as a carrier for high ammonium wastewater treatment and reported similar peaks for XRD pattern in terms of Luffa characterization.

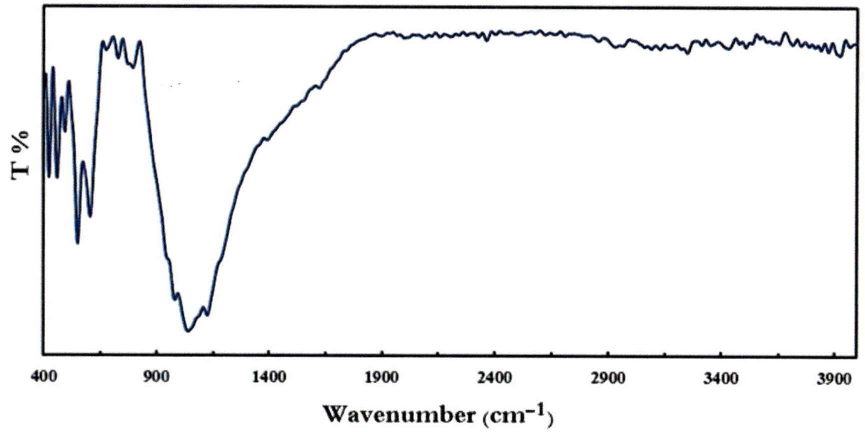

Fig. 4 The FTIR of Luffa biomaterial used for rifampin adsorption

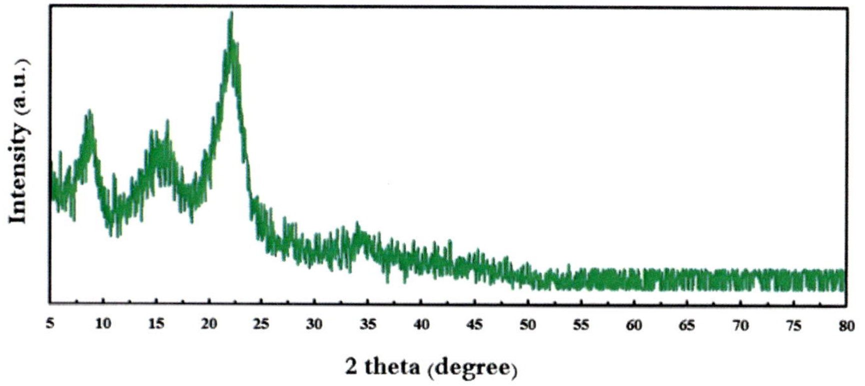

Fig. 5 XRD pattern of Luffa biomaterial used for rifampin adsorption

4.2 Optimization of the Parameters

As mentioned earlier, adsorption processes are generally affected greatly by the environmental condition, and to reach the maximum performance in removal efficiency, parameters must be optimized. In the current study concerning rifampin adsorption, parameters such as rifampin concentration, contact time, temperature, and Luffa dosage were tested. The findings are discussed in the following.

4.2.1 The Effect of Contact Time

Experiments were accomplished to measure how contact time can impact the adsorption process. Different times had been selected (0, 10, 20, 50, 80, 110, 140, and 170 min), and the purpose of these tests was to determine the removal efficiency of rifampin. The Luffa dosage and rifampin concentration were set at 0.3 gr and 30 ppm, respectively, and all experiments were conducted at 30 mL. The agitated speed was considered 100 rpm. The obtained results are illustrated in Fig. 6. The contact time can raise the removal efficiency directly, although the tests having the contact time of more than 140 min do not increase obviously. Thus, this can be learned from the tests that the optimum contact time is about 140 min. The figure shows that the removal process could be split into two stages. The first phase could be called a fast adsorption phase, and the contact time between 0–110 min showed this. The binding process that could be happened between the rifampicin and the first phase can be explained at the active adsorbent sites.

Nevertheless, it should be considered that the functional groups on the adsorbent were entirely completed, and the adsorption rate of rifampicin was regulated from the solution to the surface of the adsorbent particles. The second phase could be called the slow adsorption process, and in this phase, the adsorption rate is slowly stabilized. The adsorbent's functional groups were steadily saturated, and the adsorption rate was controlled by rifampicin transported from the external to the internal sites of adsorbent particles. The adsorption capacity value was also shown to produce similar results (Fig. 7). Briefly, there was a fast advancement in adsorption capacity until the contact time of 110 min, and then a slow trend was observed between 110–140 min. The capacity at the equilibrium level was 2.22 mg/g. Accordingly, the optimum contact time for further test was considered 140 min to achieve the maximum rifampin removal and maximum Luffa adsorption capacity. In previous studies also such behavior has been reported [34, 35].

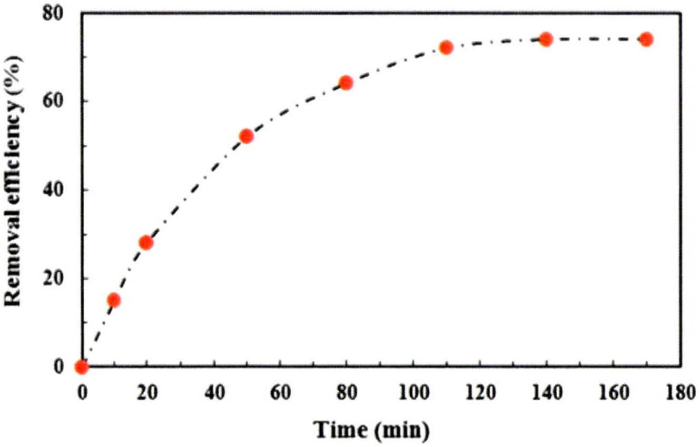

Fig. 6 The effect of contact time on rifampin removal by Luffa

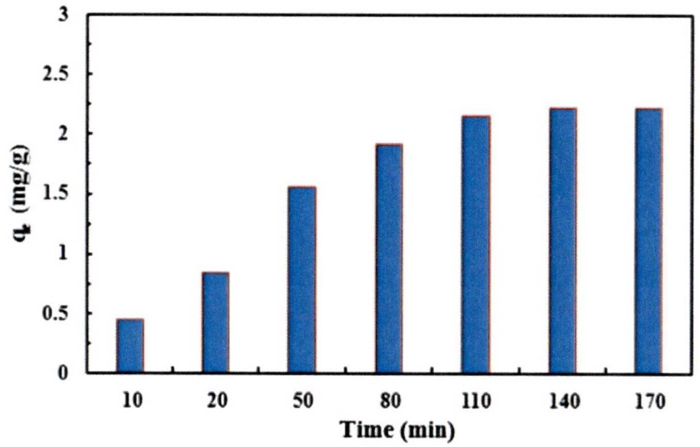

Fig. 7 The value of Luffa adsorption capacity towards rifampin molecules at different time intervals

4.2.2 The Effect of Initial Rifampin Concentration

The initial concentration of rifampin is the main adsorption parameter. Figure 8 manifests the effect of different initial rifampin concentrations ranging from 5 to 50 ppm on the removal efficiency. 0.3 gr luffa was added to 30 ml of solution, and the rotation speed and pH were adjusted to 100 rpm and 4, respectively. As it is observed, an increment in the initial rifampin concentration could decrease the removal efficiency, so the maximum removal efficiency was obtained at 5 ppm rifampin concentration, which was about 99%. The leat removal efficacy of 43% was attributed to a rifampin concentration of 50 ppm. Once the initial pollutant concentration decreased, the sites on the adsorbent are appropriate for adsorbing the molecules of rifampin [36]. In other words, high adsorption/removal efficiency is

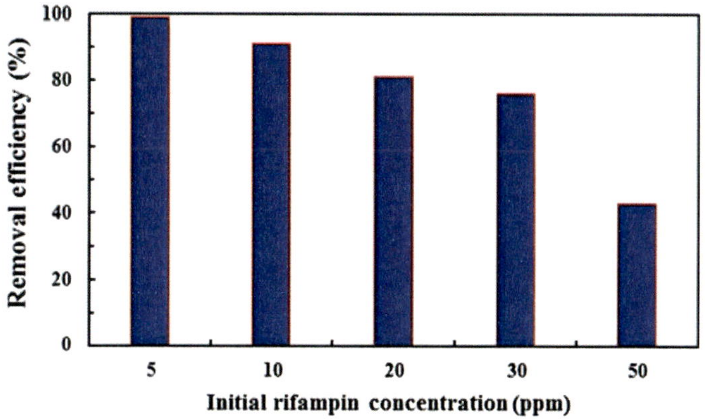

Fig. 8 The effect of the initial concentration of rifampin on the removal of rifampin by Luffa

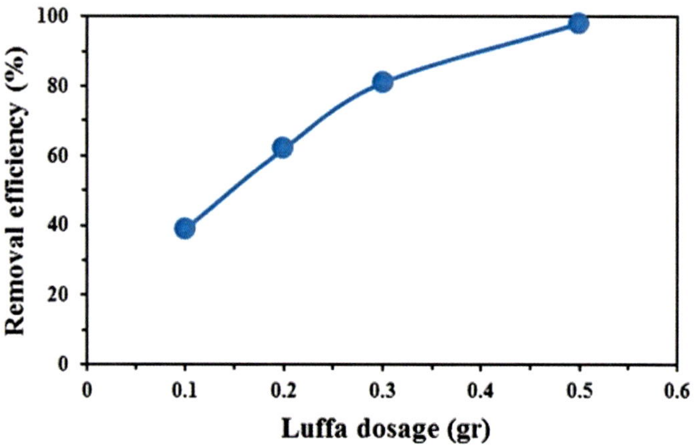

Fig. 9 Effect of the Luffa dosage on the elimination of rifampin

generally achieved at low rifampin concentration. The authors decided to continue the experiments with a rifampin concentration of 20 ppm.

4.2.3 The Effect of Luffa Dosage

Another essential element in adsorption processes is the dosage of the adsorbent. Firstly, adequate dosages of adsorbent provide required sites for pollutant removal, resulting in appropriate removal efficiency. Secondly, excessive dosages of adsorbent could generate a larger mass of wastes, which is not economically acceptable. Such optimization is carried out to minimize the treatment cost. Accordingly, the effect of Luffa dosage (0.1, 0.2, 0.3, and 0.5 gr) on rifampin removal was examined, and the findings are shown in Fig. 9. Rifampin elimination performance improved from 39 to 97% when the Luffa dose was increased from 0.1 to 0.5 gr. Sequentially, this trend is because the increasing dose of Luffa could provide more empty/functional areas for rifampin molecules' adsorbing [37]. Similar observations were reported by Mondal et al. [38], Sivaraj et al. [39], and Khadir et al. [40].

4.2.4 The Effect of Solution Temperature

Figure 10 shows how temperature can impact the removal efficiency. Rifampin elimination performance decreased from 98% to almost 80% by raising the temperature from 25 to 55 °C. Note that in temperatures 25 and 35 °C, the temperature effect was almost negligible, and somehow the process is temperature–independent. Such behavior is called exothermic processes. It may be attributed to weak interaction forces between adsorbate and adsorbent at high temperatures.

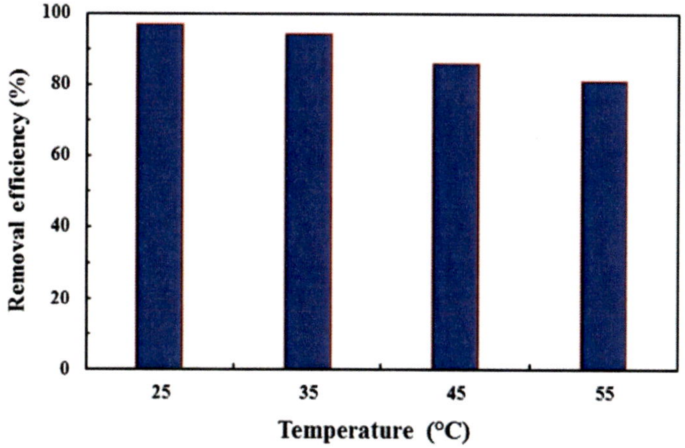

Fig. 10 The effect temperature on rifampin removal by Luff biomaterial

Based on the present literature, exothermic processes are possible once Luffa is used as the adsorbent. Raw Luffa sponge for Trypan blue dye adsorption exhibited that in 20–80 °C, the adsorption capacity has the best value at a lower temperature [41]. Khadir et al. [31] also reported an exothermic behavior of Luffa towards a non-steroidal anti-inflammatory drug.

4.3 *Isotherms of Adsorption*

Under steady temperature and pH, isotherms are absolute equations that specify the concentration of a solute on the surface of the adsorbent. Among different proposed isotherms, Langmuir, the two famous ones, Freundlich and Temkin, have been included in a wide range of investigations.

Freundlich gave an expression in 1909 describing the isothermal difference in the adsorption of an amount of gas adsorbed with pressure by a unit mass of rigid adsorbent. The Freundlich equation, also known as the Freundlich adsorption isotherm, is an experimental relationship between the concentration of a solute on the surface of an adsorbent and the concentration of the same solute in the substance with which it comes into touch. The Freundlich equation can also be obtained theoretically by attributing the change in the constant equilibrium of the binding system to the surface heterogeneity and the difference in adsorption heat. Freundlich isotherm's linear form is written as follows [42]:

$$\ln q_e = \ln K_F + \frac{1}{n} \ln C_e,$$

where

C_e (mg/L) the equilibrium concentration of rifampin;
q_e (mg/g) the equilibrium adsorption capacity;
K_F (mg/g) Freundlich adsorption isotherm constant;
n heterogeneity of adsorption process.

Figure 11 shows the linear plot of Freundlich for rifampin adsorption by Luffa, and the calculated parameters are listed in Table 1. Accordingly, the R^2 value was found 0.9547 which was an excellent value. Rifampin molecules appear to form layer (s) on the Luffa's outer surface. The n value of Freundlich isotherm is of great importance. In the current study, the n value for rifampin adsorption was 3.69, indicating a chemisorption process.

Langmuir adsorption is often used to measure and contrast the adsorptive potential of different adsorbents, which was explicitly intended to define gas–solid

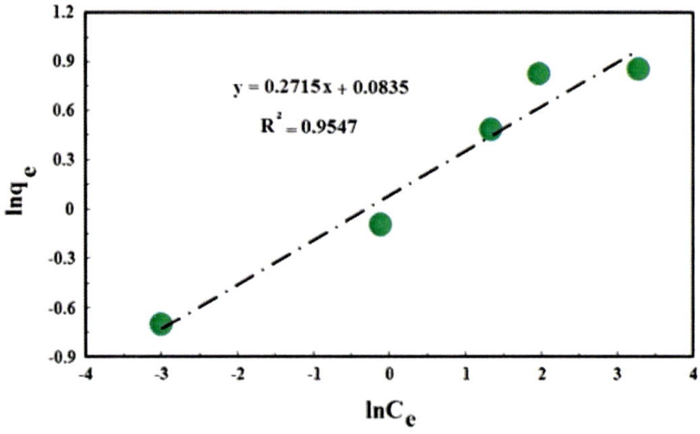

Fig. 11 The linear plot of Freundlich for rifampin adsorption by Luffa biomaterial

Table 1 The constant parameters of the studied isotherm models

Important parameters (25 °C)		Isotherm models
0.97	K_L	Langmuir
2.442	q_m	
0.9958	R^2	
1.09	K_F	Freundlich
3.69	N	
0.9547	R^2	
55.92	A_T	Temkin
0.3243	B_T	
0.8953	R^2	

phase adsorption. Adsorption and desorption, according to the model, are reversible processes. By balancing the relative adsorption and desorption rates (dynamic equilibrium), Langmuir isotherm accounts for the surface coverage. Adsorption is relative to the open surface fraction of the adsorbent, while desorption is proportional to the covered surface fraction of the adsorbent. In the following linear form, the Langmuir equation can be written [40]:

$$\frac{C_e}{q_e} = \frac{1}{q_m k_L} + \frac{C_e}{q_m},$$

where

C_e the concentration of rifampin at equilibrium (mg/g);
K_L Langmuir constant related to adsorption capacity (L/mg).

In addition, Langmuir isotherm is able to demonstrate the suitability of the adsorption process as described in the following equation with a dimensionless constant called separation factor (LD):

Crucial characteristics:

$$R_L = \frac{1}{1 + k_L c_0},$$

where C_0 is the initial rifampin concentration (mg/g).

R_L values indicate the adsorption to be

$$\text{Unfavorable} \rightarrow R_L > 1;$$

$$\text{Linear} \rightarrow R_L = 1;$$

$$\text{Favorable} \rightarrow 0 < R_L < 1;$$

$$\text{Irreversible} \rightarrow R_L = 0.$$

Figure 12 shows the linear plot of Langmuir isotherm for rifampin adsorption by Luffa, and the calculated parameters are listed in Table 1. The R^2 value of Langmuir isotherm was 0.9958 which was very close to one. It is fair to suggest that rifampin molecules tend to be adsorbed to a homogenous surface of Luffa via forming a monolayer. Based on Langmuir's assumptions, it can be stated that adsorbed rifampin on the surface of Luffa has no interaction with each other. Furthermore, the adsorption energy on the surface of Luffa is uniform and remains unchanged. The R_L values of this study are depicted in Fig. 13 demonstrating that rifampin adsorption is a favorable process.

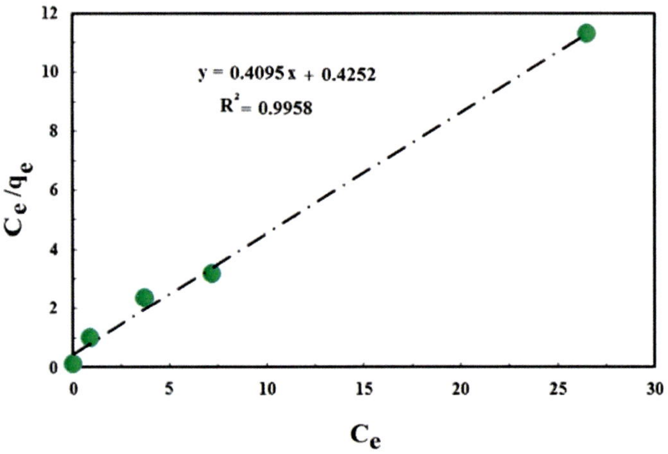

Fig. 12 The linear plot of Langmuir for rifampin adsorption by Luffa biomaterial

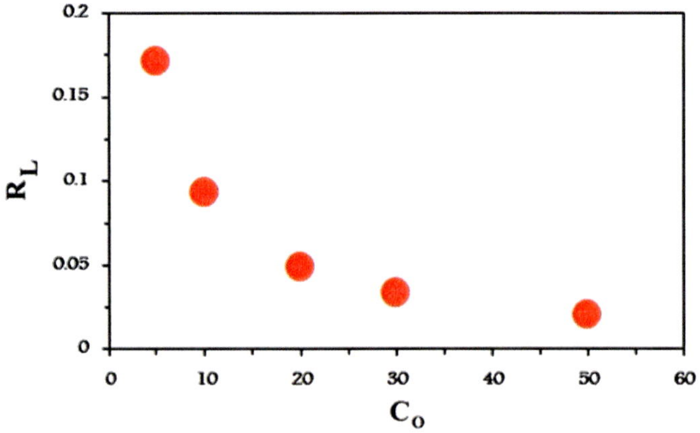

Fig. 13 Calculated R_L values for rifampin adsorption by Luffa biomaterial

The impacts of indirect adsorbate–adsorbate interactions on the adsorption mechanism are considered in the Temkin isotherm model. The Temkin isotherm believed that the heat of adsorption decreased linearly instead of logarithm when the concentration was very low or high. The Temkin linear form is given below

$$q_e = B \ln A + B \ln C_e,$$

where

B (= RT/b)	the Temkin constant related to the heat of adsorption (J/mol);
A	the Temkin isotherm equilibrium binding constant (L/g).

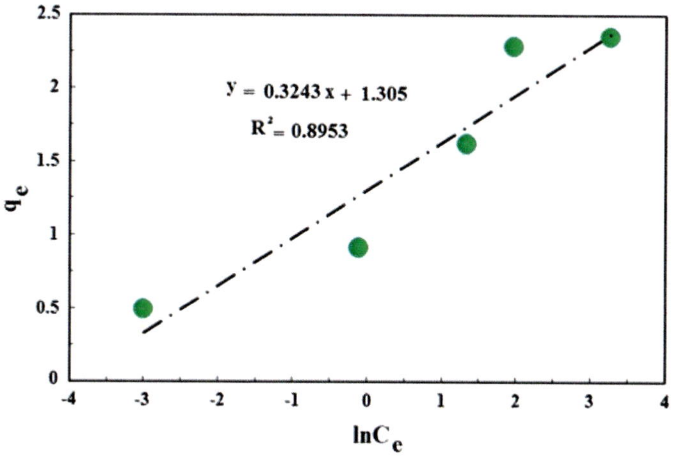

Fig. 14 The linear plot of Temkin for rifampin adsorption by Luffa biomaterial

Figure 14 shows the linear plot of Temkin isotherm for rifampin adsorption by Luffa, and the calculated parameters are listed in Table 1. Considering the very low value of R^2 (0.8953) obtained from Temkin isotherm, it is very clear that this model could not expound rifampin adsorption by Luffa.

Comparing the results of studied isotherms, the Langmuir model appears to be the best equation for explaining the pharmaceutical adsorption mechanism. A closer look at the results unveils that Freundlich also exhibited a high correlation to the experimental data in terms of R^2 values. To sum up, it can be stated that a mix of Langmuir and Freundlich assumptions would be the best model for rifampin adsorption by Luffa. This implies that the surface of Luffa might be more homogeneous with a tendency to heterogeneous in which physical and chemical sorption may occur. In the previous literature, similar results have also been reported. Jai Poinern et al. [43] combined ultrasonic/microwave methods to synthesize nanostructured hydroxyapatite for the removal of fluoride from contaminated water and reported that the equilibrium adsorption data followed both the Langmuir and Freundlich isotherms. In another study focusing on textile wastewater treatment, the results were well fitted by both Langmuir and Freundlich isotherm models [44]. Luo et al. declared that R^2 values for both Langmuir and Freundlich models were reasonable for describing the adsorptive removal of mercury by metal–organic framework material [45].

4.4 *Kinetics of Adsorption*

Lagergren proposed a pseudo-first-order kinetic model that is commonly used and accepted for solid–liquid systems and the equation as explained below

$$q_t = q_e\left(1 - \mathrm{e}^{(-K_1 \cdot t)}\right).$$

The equation that is shown above can be converted to a linear form as follows:

$$\ln(q_e - q_t) = \log q_e - K_1 t.$$

The adsorption capacity of adsorbent at equilibrium is shown by q_e (mg/g), and at the time, t (min) is q_t (mg/g), and the slope of the plot between ln $(q_e - q_t)$ versus t is K_1; it is called the rate constant (min^{-1}). For high adsorbate concentrations in the solution, the pseudo-first-order model can be effectively implemented.

In the case of low concentrated solutions, the pseudo-second-order model has the advantage of analyzing adsorption kinetics. It is presumed that chemisorption is the rate-limiting step in a pseudo-second-order reaction, and the capacity to absorb primarily depends on the adsorbent's active surface sites.

The pseudo-second-order model is expressed as follows:

$$\frac{\mathrm{d}q_t}{\mathrm{d}t} = K_2(q_e - q_t)^2.$$

The pseudo-second-order rate constant is K_2 (g/mg/min). For boundary conditions ($q_t = 0 - qt$ and $t = 0 - t$), this equation can be conveniently linearized into the equation that is mentioned below

$$\frac{t}{q_t} = \frac{1}{K_2 \cdot q_e^2} + \frac{1}{q_e} t.$$

The Elovich kinetics model is almost used in the chemical adsorption of pollutants over a heterogeneous surface. The nonlinear and linear equation of Elovich is written as follows:

$$\frac{dq_t}{dt} = \alpha \exp(-\beta q_t),$$

$$q_t = \frac{1}{\beta} Ln\, t + \frac{1}{\beta}\, Ln\, (\alpha\beta),$$

where α (mg/g min) is the initial rate constant and β (mg/g) is the desorption contact. A linear relationship must be observed in a plot of q_t versus lnt.

Figure 15 shows the linear plots of the pseudo-first-order, pseudo-second-order, and Elovich models, and their relative parameters are summarized in Table 2. Comparing the kinetics models in terms of R^2 value, it shows that Elovich and pseudo-second-order models have the highest value which is very close to unity. Pseudo-first-order could not describe the kinetics of rifampin adsorption very well which is in great conformity with many studies in adsorption systems that believed this kinetic model is not the excellent one [46, 47]. Therefore, it should be noted that

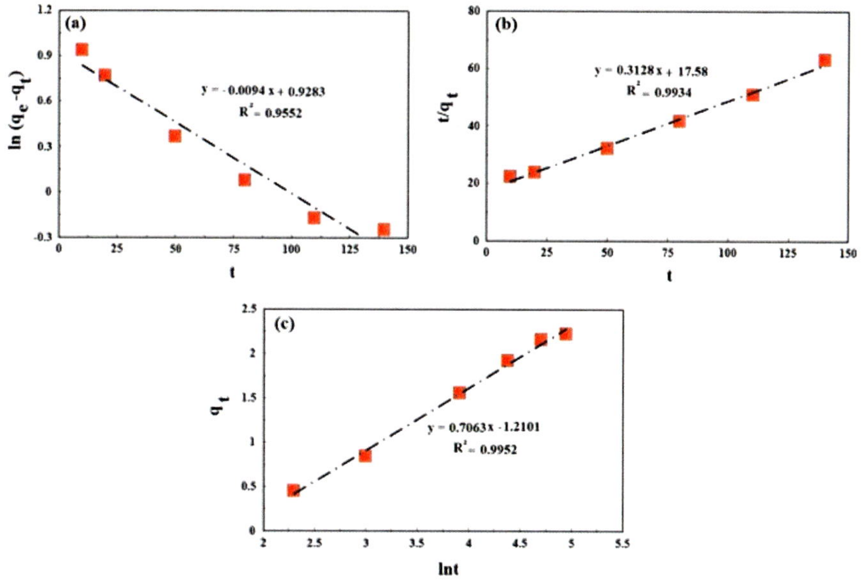

Fig. 15 Linear kinetics models of pseudo-first-order model (**a**), pseudo-second-order model (**b**), and Elovich (**c**) for rifampin adsorption by Luffa biomaterial

Table 2 The constant parameters of the studied kinetics models

Important parameters		Kinetics models
0.009	K_1	PFO
2.53	q_e	
0.9552	R^2	
0.017	K_2	PSO
3.196	q_e	
0.9934	R^2	
3.92	α	Elovich
1.415	β	
0.9952	R^2	

the Luffa surface might be energetically heterogeneous towards rifampin molecules, and chemical adsorption or chemisorption is the rate-limiting step. Lin et al. [14] reported that iron nanoparticles were used to remove Pb (II) and rifampin from a polluted solution, and the kinetics mechanism was found to adopt pseudo-second-order kinetics. Not only rifampin, other pharmaceutical removal kinetics followed pseudo-second-order or Elovich, including Khadir et al. [48] and Sekulic et al. [49].

4.5 Conclusion

Water pollution is recognized as a universal problem nowadays, and many attempts are being done to mitigate the level of contamination. The present work focuses on the elimination of colouring agents produced from rifampin drug in waters. Luffa was used as received without further modification. Increased contact time was beneficial for both removal efficiency and adsorption capacity. It was found that higher dosages of Luffa resulted in greater rifampin removal which may be attributed to the enhancement in the adsorbent sites. The whole process followed the assumptions of Langmuir, pseudo-second-order and Elovich models. High removal efficiency achieved in this study proved that Luffa biomaterial is a sustainable, non-toxic and cost-effective biomaterial with great availability in nature that can effectively mitigate and remove contaminants in aquatic media.

Declaration of Competing Interest The authors declare that they have no known competing interests or personal relationships that could have appeared to influence the work reported in this paper.

References

1. Khadir A, Mollahosseini A, Tehrani RMA, Negarestani M (2020) A review on pharmaceutical removal from aquatic media by adsorption: understanding the influential parameters and novel adsorbents BT—sustainable green chemical processes and their allied applications. In: Inamuddin AA (eds) Springer International Publishing, Cham, pp 207–265. https://doi.org/10.1007/978-3-030-42284-4_8
2. Khadir A, Mollahosseini A, Negarestani M, Mardy A (2020) Anaerobic Biotechnology for the treatment of pharmaceutical compounds and hospital wastewaters BT—methods for bioremediation of water and wastewater pollution. In: Inamuddin, Ahamed MI, Lichtfouse E, Asiri AM (eds) Springer International Publishing, Cham 61–84. https://doi.org/10.1007/978-3-030-48985-4_3
3. Danner M-C, Robertson A, Behrends V, Reiss J (2019) Antibiotic pollution in surface fresh waters: occurrence and effects. Sci Total Environ 664:793–804. https://doi.org/10.1016/j.scitotenv.2019.01.406
4. Negarestani M, Motamedi M, Kashtiaray A, Khadir A, Sillanpää M (2020) Simultaneous removal of acetaminophen and ibuprofen from underground water by an electrocoagulation unit: operational parameters and kinetics. Groundw Sustain Dev 11:100474. https://doi.org/10.1016/j.gsd.2020.100474
5. Mollahosseini A, Khadir A, Saeidian J (2019) Core–shell polypyrrole/Fe_3O_4 nanocomposite as sorbent for magnetic dispersive solid-phase extraction of Al+3 ions from solutions: investigation of the operational parameters. J Water Process Eng 29:100795. https://doi.org/10.1016/j.jwpe.2019.100795
6. Ghenaatgar A, Tehrani RMA, Khadir A (2019) Photocatalytic degradation and mineralization of dexamethasone using WO3 and ZrO2 nanoparticles: optimization of operational parameters and kinetic studies. J Water Process Eng 32:100969. https://doi.org/10.1016/j.jwpe.2019.100969

7. Rivera-Utrilla J, Sánchez-Polo M, Ferro-García MÁ, Prados-Joya G, Ocampo-Pérez R (2013) Pharmaceuticals as emerging contaminants and their removal from water. A Rev Chemosph 93:1268–1287. https://doi.org/10.1016/j.chemosphere.2013.07.059
8. Velempini T, Prabakaran E, Pillay K (2021) Recent developments in the use of metal oxides for photocatalytic degradation of pharmaceutical pollutants in water—a review. Mater Today Chem 19:100380. https://doi.org/10.1016/j.mtchem.2020.100380
9. Huang L, Shen R, Shuai Q (2021) Adsorptive removal of pharmaceuticals from water using metal-organic frameworks: a review. J Environ Manage 277:111389. https://doi.org/10.1016/j.jenvman.2020.111389
10. Pérez-Ibarbia L, Majdanski T, Schubert S, Windhab N, Schubert US (2016) Safety and regulatory review of dyes commonly used as excipients in pharmaceutical and nutraceutical applications. Eur J Pharm Sci 93:264–273. https://doi.org/10.1016/j.ejps.2016.08.026
11. Allam KV, Kumar GP (2011) Colorants—the cosmetics for the pharmaceutical dosage forms. Int J Pharm Pharm Sci 3:13–21
12. Henrique DC, Quintela DU, Ide AH, Erto A, Duarte JLDS, Meili L (2020) Calcined Mytella falcata shells as alternative adsorbent for efficient removal of rifampicin antibiotic from aqueous solutions. J Environ Chem Eng. https://doi.org/10.1016/j.jece.2020.103782
13. Cai W, Weng X, Chen Z (2019) Highly efficient removal of antibiotic rifampicin from aqueous solution using green synthesis of recyclable nano-Fe_3O_4. Environ Pollut. https://doi.org/10.1016/j.envpol.2019.01.108
14. Lin Z, Weng X, Owens G, Chen Z (2020) Simultaneous removal of Pb(II) and rifampicin from wastewater by iron nanoparticles synthesized by a tea extract. J Clean Prod. https://doi.org/10.1016/j.jclepro.2019.118476
15. Liu L, Xu Q, Owens G, Chen Z (2021) Fenton-oxidation of rifampicin via a green synthesized rGO@nFe/Pd nanocomposite. J Hazard Mater 402:123544. https://doi.org/10.1016/j.jhazmat.2020.123544
16. Madivoli ES, Kareru PG, Makhanu DS, Wandera KS, Maina EG, Wanakai SI, Kimani PK (2020) Synthesis of spherical titanium dioxide microspheres and its application to degrade rifampicin. Environ Nanotechnol Monit Manag. https://doi.org/10.1016/j.enmm.2020.100327
17. Xu Q, Owens G, Chen Z (2020) Adsorption and catalytic reduction of rifampicin in wastewaters using hybrid rGO@Fe/Pd nanoparticles. J Clean Prod 264:121617. https://doi.org/10.1016/j.jclepro.2020.121617
18. da Silva Duarte JL, Solano AMS, Arguelho MLPM, Tonholo J, Martínez-Huitle CA, de Paiva e Silva CL (2018) Evaluation of treatment of effluents contaminated with rifampicin by Fenton, electrochemical and associated processes. J Water Process Eng 22:250–257. https://doi.org/10.1016/j.jwpe.2018.02.012
19. Khadir A, Motamedi M, Pakzad E, Sillanpää M, Mahajan S (2020) The prospective utilization of Luffa fibres as a lignocellulosic bio-material for environmental remediation of aqueous media: a review. J Environ Chem Eng 104691. https://doi.org/10.1016/j.jece.2020.104691
20. Ad C, Djedid M, Benalia M, Boudaoud A, Elmsellem H, Ben Saffedine F (2018) Adsorptive Removal of Nickel (II) Using Luffa cylindrica: effect of NaCl concentration on equilibrium and kinetic parameters BT—recent advances in environmental science from the euro-mediterranean and surrounding regions. In: Kallel A, Ksibi M, Ben Dhia H, Khélifi N (eds) Springer International Publishing, Cham, pp 1305–1306
21. Xiao F, Cheng J, Cao W, Yang C, Chen J, Luo Z (2019) Removal of heavy metals from aqueous solution using chitosan-combined magnetic biochars. J Colloid Interface Sci 540:579–584. https://doi.org/10.1016/j.jcis.2019.01.068
22. Zeng L, Liu Q, Lu M, Liang E, Wang G, Xu W (2019) Modified natural loofah sponge as an effective heavy metal ion adsorbent: amidoxime functionalized poly(acrylonitrile-g-loofah). Chem Eng Res Des. https://doi.org/10.1016/j.cherd.2019.07.021
23. Kong Q, Wang Y, Shu L, Miao M (2016) Isotherm, kinetic, and thermodynamic equations for cefalexin removal from liquids using activated carbon synthesized from loofah sponge. Desalin Water Treat 57:7933–7942. https://doi.org/10.1080/19443994.2015.1052991

24. Khadir A, Negarestani M, Mollahosseini A (2020) Sequestration of a non-steroidal anti-inflammatory drug from aquatic media by lignocellulosic material (Luffa cylindrica) reinforced with polypyrrole: study of parameters, kinetics, and equilibrium. J Environ Chem Eng. Elsevier
25. Yu J, Wang L, Chi R, Zhang Y, Xu Z, Guo J (2013) Removal of cationic dyes: basic magenta and methylene blue from aqueous solution by adsorption on modified loofah. Res Chem Intermed 39:3775–3790. https://doi.org/10.1007/s11164-012-0880-7
26. Caicedo O, Devia-Ramirez J, Malagón A (2018) Adsorption of common laboratory dyes using natural fibers from Luffa cylindrica. J Chem Educ 95:2233–2237. https://doi.org/10.1021/acs.jchemed.8b00156
27. Mashkoor F, Nasar A (2019) Preparation, characterization and adsorption studies of the chemically modified Luffa aegyptica peel as a potential adsorbent for the removal of malachite green from aqueous solution. J Mol Liq 274:315–327. https://doi.org/10.1016/j.molliq.2018.10.119
28. Demir H, Top A, Balköse D, Ülkü S (2008) Dye adsorption behavior of Luffa cylindrica fibers. J Hazard Mater 153:389–394. https://doi.org/10.1016/j.jhazmat.2007.08.070
29. El Ashtoukhy ESZ (2009) Loofa egyptiaca as a novel adsorbent for removal of direct blue dye from aqueous solution. J Environ Manage 90:2755–2761. https://doi.org/10.1016/j.jenvman.2009.03.005
30. Petit S, Madejova J (2013) Chapter 2.7—Fourier transform infrared spectroscopy. Bergaya F, Lagaly CS (eds) Handb Clay Sci. Elsevier, pp 213–231. https://doi.org/10.1016/B978-0-08-098259-5.00009-3
31. Khadir A, Negarestani M, Mollahosseini A (2020) Sequestration of a non-steroidal anti-inflammatory drug from aquatic media by lignocellulosic material (Luffa cylindrica) reinforced with polypyrrole: study of parameters, kinetics, and equilibrium. J Environ Chem Eng 103734. https://doi.org/10.1016/j.jece.2020.103734
32. Boudechiche N, Mokaddem H, Sadaoui Z, Trari M (2016) Biosorption of cationic dye from aqueous solutions onto lignocellulosic biomass (Luffa cylindrica): characterization, equilibrium, kinetic and thermodynamic studies. Int J Ind Chem 7:167–180. https://doi.org/10.1007/s40090-015-0066-4
33. Zhang J, Yang J, Tian Q, Liang X, Zhu Y, Sand W, Li F, Ma C, Liu Y, Yang B (2019) Durability and performance of loofah sponge as carrier for wastewater treatment with high ammonium. Water Environ Res 91:581–587. https://doi.org/10.1002/wer.1067
34. Mirjavadi ES, Tehrani RMA, Khadir A (2019) Effective adsorption of zinc on magnetic nanocomposite of Fe_3O_4/zeolite/cellulose nanofibers: kinetic, equilibrium, and thermodynamic study. Environ Sci Pollut Res. https://doi.org/10.1007/s11356-019-06165-z
35. Pahlavanzadeh H, Motamedi M (2020) Adsorption of Nickel, Ni(II), in aqueous solution by modified zeolite as a cation-exchange adsorbent. J Chem Eng Data 65:185–197. https://doi.org/10.1021/acs.jced.9b00868
36. dos Santos RMM, Gonçalves RGL, Constantino VRL, Santilli CV, Borges PD, Tronto J, Pinto FG (2017) Adsorption of Acid Yellow 42 dye on calcined layered double hydroxide: Effect of time, concentration, pH and temperature. Appl Clay Sci 140:132–139. https://doi.org/10.1016/j.clay.2017.02.005
37. Yadav AK, Kaushik CP, Haritash AK, Kansal A, Rani N (2006) Defluoridation of groundwater using brick powder as an adsorbent. J Hazard Mater 128:289–293
38. Mondal P, Majumder CB, Mohanty B (2008) Effects of adsorbent dose, its particle size and initial arsenic concentration on the removal of arsenic, iron and manganese from simulated ground water by Fe3+ impregnated activated carbon. J Hazard Mater 150:695–702. https://doi.org/10.1016/j.jhazmat.2007.05.040
39. Sivaraj R, Namasivayam C, Kadirvelu K (2001) Orange peel as an adsorbent in the removal of Acid violet 17 (acid dye) from aqueous solutions. Waste Manag 21:105–110. https://doi.org/10.1016/S0956-053X(00)00076-3
40. Khadir A, Negarestani M, Ghiasinejad H (2020) Low-cost sisal fibers/polypyrrole/polyaniline biosorbent for sequestration of reactive orange 5 from aqueous solutions. J Environ Chem Eng 8:103956. https://doi.org/10.1016/j.jece.2020.103956

41. Nadaroglu H, Cicek S, Gungor AA (2017) Removing Trypan blue dye using nano-Zn modified Luffa sponge, Spectrochim. Acta Part A Mol Biomol Spectrosc 172:2–8. https://doi.org/10.1016/j.saa.2016.08.052
42. Yuan X, Wu Z, Zhong H, Wang H, Chen X, Leng L, Jiang L, Xiao Z, Zeng G (2016) Fast removal of tetracycline from wastewater by reduced graphene oxide prepared via microwave-assisted ethylenediamine–N, N′–disuccinic acid induction method. Environ Sci Pollut Res 23:18657–18671. https://doi.org/10.1007/s11356-016-6892-x
43. Poinern GEJ, Ghosh MK, Ng Y-J, Issa, Anand S, Singh P (2011) Defluoridation behavior of nanostructured hydroxyapatite synthesized through an ultrasonic and microwave combined technique. J Hazard Mater 185:29–37. https://doi.org/10.1016/j.jhazmat.2010.08.087
44. Djehaf K, Bouyakoub AZ, Ouhib R, Benmansour H, Bentouaf A, Mahdad A, Moulay N, Bensaid D, Ameri M (2017) Textile wastewater in Tlemcen (Western Algeria): impact, treatment by combined process. Chem Int 3:314–318
45. Luo F, Chen JL, Dang LL, Zhou WN, Lin HL, Li JQ, Liu SJ, Luo MB (2015) High-performance Hg2+ removal from ultra-low-concentration aqueous solution using both acylamide- and hydroxyl-functionalized metal–organic framework. J Mater Chem A 3:9616–9620. https://doi.org/10.1039/C5TA01669J
46. Kebede TG, Dube S, Nindi MM (2019) Biopolymer electrospun nanofibres for the adsorption of pharmaceuticals from water systems. J Environ Chem Eng 7:103330. https://doi.org/10.1016/j.jece.2019.103330
47. Martini BK, Daniel TG, Corazza MZ, de Carvalho AE (2018) Methyl orange and tartrazine yellow adsorption on activated carbon prepared from boiler residue: kinetics, isotherms, thermodynamics studies and material characterization. J Environ Chem Eng 6:6669–6679. https://doi.org/10.1016/j.jece.2018.10.013
48. Khadir A, Motamedi M, Negarestani M, Sillanpää M, Sasani M (2020) Preparation of a nano bio-composite based on cellulosic biomass and conducting polymeric nanoparticles for ibuprofen removal: kinetics, isotherms, and energy site distribution. Int J Biol Macromol https://doi.org/10.1016/j.ijbiomac.2020.06.095
49. Turk Sekulic M, Boskovic N, Milanovic M, Grujic Letic N, Gligoric E, Pap S (2019) An insight into the adsorption of three emerging pharmaceutical contaminants on multifunctional carbonous adsorbent: mechanisms, modelling and metal coadsorption. J Mol Liq 284:372–382. https://doi.org/10.1016/j.molliq.2019.04.020